TIME

100 New Scientific Discoveries
Fascinating, Momentous, and Mind-Expanding Stories

PRINT YOUR OWN *3-D printers allow anyone to produce objects like this lampshade, as well as toys, jewelry, or ... pork chops?*

TIME

MANAGING EDITOR Richard Stengel
DESIGN DIRECTOR D.W. Pine
DIRECTOR OF PHOTOGRAPHY Kira Pollack

100 New Scientific Discoveries

EDITOR Stephen Koepp
DESIGNER Sharon Okamoto
PHOTO EDITOR Dot McMahon
ASSISTANT PHOTO EDITOR Richard Boeth
WRITERS David Bjerklie, Michael Q. Bullerdick, Jeffrey Kluger, Michael Lemonick, Alice Park, Ellen Shapiro, Bryan Walsh
COPY EDITOR David Olivenbaum
REPORTERS Elizabeth Bland, Jenisha Watts
EDITORIAL PRODUCTION Richard Prue, Lionel P. Vargas, David Sloan
GRAPHICS EDITOR Lon Tweeten

Time Home Entertainment

PUBLISHER Jim Childs
VICE PRESIDENT, BUSINESS DEVELOPMENT & STRATEGY Steven Sandonato
EXECUTIVE DIRECTOR, MARKETING SERVICES Carol Pittard
EXECUTIVE DIRECTOR, RETAIL & SPECIAL SALES Tom Mifsud
EXECUTIVE PUBLISHING DIRECTOR Joy Butts
DIRECTOR, BOOKAZINE DEVELOPMENT & MARKETING Laura Adam
FINANCE DIRECTOR Glenn Buonocore
ASSOCIATE PUBLISHING DIRECTOR Megan Pearlman
ASSISTANT GENERAL COUNSEL Helen Wan
ASSISTANT DIRECTOR, SPECIAL SALES Ilene Schreider
BOOK PRODUCTION MANAGER Suzanne Janso
DESIGN & PREPRESS MANAGER Anne-Michelle Gallero
BRAND MANAGER Michela Wilde
ASSOCIATE BRAND MANAGER Isata Yansaneh
ASSOCIATE PREPRESS MANAGER Alex Voznesenskiy

EDITORIAL DIRECTOR Stephen Koepp
EDITORIAL OPERATIONS DIRECTOR Michael Q. Bullerdick

SPECIAL THANKS TO: Katherine Barnet, Jeremy Biloon, Stephanie Braga, Susan Chodakiewicz, Rose Cirrincione, Lauren Hall Clark, Jacqueline Fitzgerald, Christine Font, Jenna Goldberg, Hillary Hirsch, David Kahn, Amy Mangus, Robert Marasco, Kimberly Marshall, Amy Migliaccio, Nina Mistry, Dave Rozzelle, Ricardo Santiago, Adriana Tierno, Vanessa Wu, TIME Imaging

ISBN 10: 1-61893-076-1
ISBN 13: 978-1-61893-076-7
Library of Congress Control Number: 2012951275

We welcome your comments and suggestions about TIME Books. Please write to us at TIME Books, Attention: Book Editors, P.O. Box 11016, Des Moines, IA 50336-1016.
If you would like to order any of our hardcover Collector's Edition books, please call us at 1-800-327-6388, Monday through Friday, 7 a.m. to 8 p.m., or Saturday, 7 a.m. to 6 p.m., Central Time.

Some of the articles in this book were previously published in substantially the same form in TIME magazine or on Time.com from 2011 through 2012.

HIGGS HUNTER *Experiments at the Large Hadron Collider led to the discovery of the Higgs boson, the elusive particle that explains why mass exists in the universe.*

Contents

MELTING AWAY *Decaying fragments of Iceland's mammoth Vatnajökull glacier are part of a trend of shrinking sea ice that goes back at least 30 years.*

JAMES BALOG/AURORA PHOTOS

The Beautiful Blizzard of Science

By Jeffrey Kluger

Jonas Salk didn't doodle—or at least he didn't doodle much. It's not as if Salk had much time to doodle, not when he was in a constant foot race with the calendar, trying to develop a vaccine against polio before the fever season of summer would arrive, paralyzing or killing tens of thousands of children before the first chill of fall. But still, the almost entire lack of any doodles in Salk's personal papers surprised me.

I spent the better part of a year wading through those papers, tens of thousands of them, stored in a climate-controlled vault at the University of California, San Diego, while I was researching a book about Salk. There were office memos and minutes of meetings and lab notebooks filled with page upon page of pointillist data, much in Salk's own hands, and in all of it I saw just a single, stray scribble—a little triangular thing, absently drawn and surely instantly forgotten.

That, in a way, is how it should be. We live in a world that craves the eureka moment—the something-for-nothing insight or discovery that opens up whole new horizons at a stroke. Once in a great while that happens. But typically, science is a much slower thing than that: an inch-by-inch process in which data point is stacked atop data point, micro-insight atop micro-insight, until a weight suddenly shifts, a balance suddenly tips, and the world changes. That leaves no room for doodles or distractions. It leaves room only for the notebooks and the work.

Those big breakthroughs don't happen often, though in the year just past there were several. The Higgs boson, the elementary particle that holds together the very edifice of Einsteinian theory, was found. The Mars Curiosity rover, the most sophisticated unmanned space probe ever built, set its wheels gently in the Martian soil. And all over the other scientific disciplines—chemistry, zoology, medicine, psychology, archaeology—there were triumphs big and small. Some of them, like recent breakthroughs in stem-cell research, will bring healing. Some of them, like the discovery of exoplanets orbiting other stars, may reveal extraterrestrial life. Some of them will help us save our own ravaged planet, or protect our threatened animals, or understand our evolutionary past. All of them will leave us better, healthier, or just plain smarter than we were before.

Science does move slowly, but only until—like an avalanche built from a trillion, trillion flakes—it moves very fast. This year, like all years, those snows shifted again.

The Cosmos

Curiosity Meets Mars

*A one-ton rover can teach us a lot about
the red planet—and the blue one too.*

By Jeffrey Kluger

The folks in mission control at NASA's Jet Propulsion Laboratory ate a lot of peanuts in the minutes leading up to the landing of the Curiosity rover on Mars. Peanuts have been the order of the day at JPL when a spacecraft is preparing to land ever since July 31, 1964, when the Ranger 7 probe was making its final approach to the moon. Ranger's job was a simple one: crash-land on the lunar surface, snapping pictures to beam home on the way down. But Rangers 1 through 6 had failed to do even that, and the JPL engineers knew they were about out of chances. Ranger 7 at last broke that losing streak—and, as it happened, someone was nibbling peanuts during the landing. That, the missile men of JPL figured, must have been a good-luck charm. No one has dared defy it since.

But it would take more than luck and peanuts to get Curiosity safely to the surface of Mars. At 1:25 a.m. ET on August 6, 2012, the SUV-size rover, sealed inside a blunt-bottomed capsule, would slam into the Martian atmosphere at a blazing 13,000 mph. Seven miles above the surface, when the thin air had slowed the ship to 900 mph, its heat shield would pop away, and it would deploy a billowing parachute. Its retrorockets would then bring the rover and its housing to a near hover just two stories above the surface, where it would be lowered to the ground by wire cables—a $2.5 billion extraterrestrial marionette, settling its wheels gently into the red soil.

It all could have gone very wrong—but when the time finally came, people watching NASA's live feed saw a flight-dynamics engineer call out, "Stand by for sky crane." And then, less than a minute later: "Touchdown confirmed! We're safe on Mars!" This was

much more than just another triumph for NASA, however. In an era in which the grind and gridlock of Washington have made citizens wary of anything the government touches, this was a reminder of what the nation can still do.

The scene in mission control was what smart looks like. It was what vision looks like. And it's not just Curiosity: a country that can't get its roads and bridges fixed at home actually has infrastructure on Mars. Two NASA orbiters—Mars Global Surveyor and Mars Odyssey—helped relay Curiosity's transmissions to Earth and wave their new sister in for her landing. And even as Curiosity settles down to work, no fewer than eight other NASA probes are ranging through the solar system, exploring—or on their way to explore—the moon, Mercury, Jupiter, Saturn, Pluto, the asteroid Ceres, and the interstellar void beyond the planets.

It's Curiosity, however, that defines the state of the exploratory art. With 10 instruments weighing a collective 15 times more than those aboard previous rovers Spirit and Opportunity, it will study the geology, chemistry, and possible biology of Mars, looking for signs of carbon, methane, and other organic fingerprints on a world that a few billion years ago was warm and fairly sloshing with water—the most basic requirement for life. Earlier probes have strongly made the case that Martian life, either extant or ancient, is possible. Curiosity could seal the deal. "We all feel a sense of pressure to do something profound," says geologist and project scientist John Grotzinger.

In some ways, they've already done that by framing an unavoidable question: if we can do this exceedingly hard thing so well, why do we make such a hash of the challenges at home, the inventing and investing that 21st-century progress demands? Help answer that, and Curiosity could achieve great things on two worlds at once.

On Mars, the achievement will be purely scientific. The Red Planet may be a meteor-blasted desert today, but its surface is marked with what look uncannily like dry riverbeds and the dusty beds of desiccated oceans. Up close, the Spirit and Opportunity rovers uncovered a wealth of minerals that generally form in the presence of water. Mars was wet for only a billion or so of its 4.5 billion years, but as Earth's history proves, that's plenty of time to cook up life.

Curiosity's landing site was a formation known as Gale Crater, 96 miles wide, located in Mars's southern hemisphere. It's thought to be up to 3.8 billion years old, well within the time when Mars was probably wet, meaning the crater was once a large lake. Three-mile-high Mount Sharp rises at its center, with exposed strata layer-caked down its sides. Channels that appear to have been carved by water run down both the crater walls and the mountain base. Radiating channels that look like the river deltas on Earth mark the soil near the prime landing site. All this is irresistible to geologists searching for the basic conditions for life.

Curiosity is conducting that search in many ways. The rover's arm will scoop up samples of soil and deliver them to an onboard analysis chamber, where they will be studied by a gas chromatograph, a mass spectrometer, and a laser spectrometer, looking for telltale isotopes. Chemical sniffers will sample the Martian air for carbon compounds—especially methane—which are the building blocks and byproducts of life. A million-watt laser will vaporize rocks up to 23 feet away, allowing a spectrometer to analyze

> **6 July 12**
> **Zoom! I'm speeding toward Mars at nearly 48,000 mph relative to the sun. Countdown to landing: 30 days.**
>
> —Curiosity's Twitter feed, @MarsCuriosity, written by social-media specialists

Decades of Discovery: Mars and Beyond

Phoenix
ACTIVE
5 MONTHS

OLYMPUS MONS

Viking 1
6 YEARS
3 MONTHS

THARSIS MONTES

VALLES MARINERI

Opportunity
8 YEARS 10 MONTHS
STILL ACTIVE*

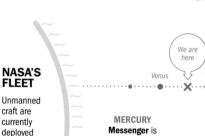

TOUCHDOWN ON THE RED PLANET

Successful landing missions

By launch date

1960 · · · · · · · · · · 1970

| Failed landing missions | Sputnik 24 *USSR* Failed to leave Earth's orbit | Mars 2 *USSR* Crashed on Mars' surface | Mars *USSR* Landed s then d |

* AS OF NOVEMBER 2012.

1

NASA'S FLEET

Unmanned craft are currently deployed throughout the solar system

We are here

Venus

MERCURY
Messenger is studying the tiny planet's surface and magnetic field and has found volcanic deposits

LON TWEETEN AND HEATHER JONES

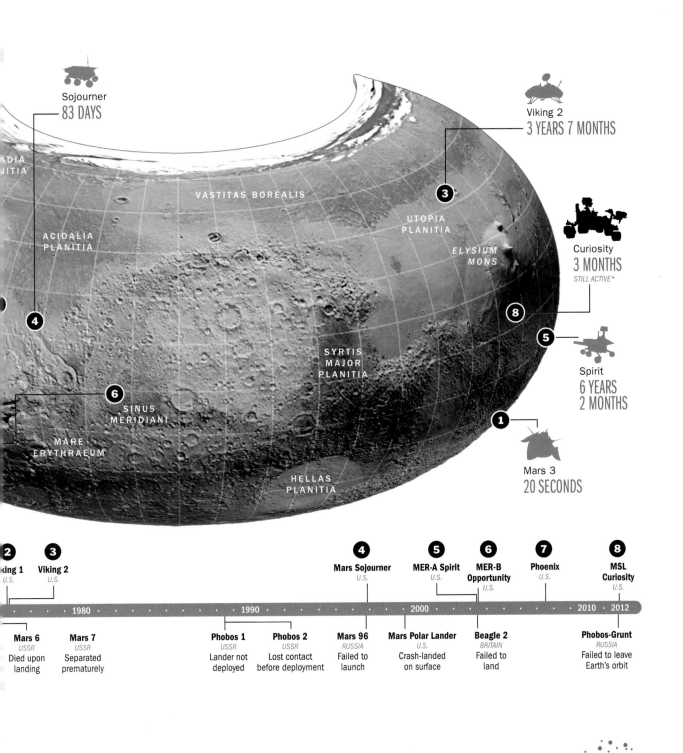

Sojourner
83 DAYS

Viking 2
3 YEARS 7 MONTHS

VASTITAS BOREALIS

ADIA
ITIA

ACIDALIA
PLANITIA

UTOPIA
PLANITIA

ELYSIUM
MONS

Curiosity
3 MONTHS
STILL ACTIVE*

SYRTIS
MAJOR
PLANITIA

Spirit
6 YEARS
2 MONTHS

SINUS
MERIDIANI

MARE
ERYTHRAEUM

HELLAS
PLANITIA

Mars 3
20 SECONDS

2 king 1	**3** Viking 2		**4** Mars Sojourner	**5** MER-A Spirit	**6** MER-B Opportunity	**7** Phoenix	**8** MSL Curiosity
U.S.	U.S.		U.S.	U.S.	U.S.	U.S.	U.S.

1980 · · · 1990 · · · 2000 · · · 2010 · 2012

| **Mars 6** USSR Died upon landing | **Mars 7** USSR Separated prematurely | **Phobos 1** USSR Lander not deployed | **Phobos 2** USSR Lost contact before deployment | **Mars 96** RUSSIA Failed to launch | **Mars Polar Lander** U.S. Crash-landed on surface | **Beagle 2** BRITAIN Failed to land | **Phobos-Grunt** RUSSIA Failed to leave Earth's orbit |

MARS
Multiple active spacecraft including Curiosity, Opportunity, Mars Global Surveyor and Mars Odyssey study the planet

ASTEROID VESTA AND DWARF PLANET CERES
Dawn is investigating the asteriod belt's two largest objects to learn more about the early days of the solar system

JUPITER
Juno will peer underneath dense clouds to learn about the planet's chemical makeup and its turbulent atmosphere, including the storm known as the Great Red Spot

SATURN
Cassini has revealed saltwater geysers on the moon Enceladus (which could contain organic material), and its observations of the moon Titan suggest what Earth might have been like before life emerged

VOYAGER 1 AND 2
Voyager 1 flew by Jupiter and Saturn; **Voyager 2** reconnoitered those worlds as well as Uranus and Neptune. Both probes are now leaving the solar system for interstellar space

PLUTO AND THE KUIPER BELT
New Horizons will study Pluto, its moons and the swarm of icy, comet-like bodies that surround the solar system

11

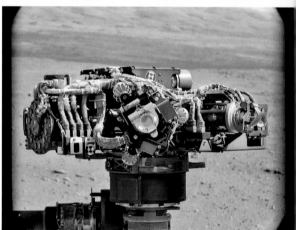

the chemistry of the residue. An X-ray spectrometer will do similar work on rocks near the rover.

Most appealing for the folks back home are the 17 cameras arrayed around Curiosity. They have the visual acuity to resolve an object the size of a golf ball 27 yards away and the resolution to capture one-megapixel color images from multiple perspectives. The sharpest of these imagers is mounted atop the rover's vertical mast, which, extended, rises seven feet above ground. "You could not look this thing in the eye unless you were an NBA player," says mission-systems manager Mike Watkins.

President Obama was quick to make hay out of good news from space. "Tonight, on the planet Mars, the United States of America made history," he said in an official statement. "It proves that even the longest odds are no match for our unique blend of ingenuity and determination." And NASA administrator Charles Bolden Jr. was quick to give props to the president who appointed him. "President Obama has laid out a bold vision for sending humans to Mars in the mid-2030s," he said, "and today's landing marks a significant step in achieving this goal." But Obama's record on space has been mixed, especially when it comes to human exploration. The last shuttle flew in July 2011, but its successor, a heavy-lift booster called the Space

Self-Portraits and Landscapes

Curiosity's cameras have taken the sharpest images ever seen of the Martian surface, and documented its own presence as well. One self-portrait (left) shows the rover's deck and two of its six wheels; another (above) shows a microscopic camera on Curiosity's robot arm, and a third (top, far right) captures tracks from the craft's first test drive. A close-up (this page, top) reveals rocks that formed in water, while the shot to its right looks out to the layered slopes of Mount Sharp, inside Gale Crater, where Curiosity landed; at far right, Mount Sharp is seen from another angle.

Launch System, coupled to a crew vehicle called Orion, won't take astronauts into space until 2021. After that, it will fly every other year at best, and it's not clear where the astronauts will go—to an asteroid, maybe, or maybe to Mars. "This is a pace that doesn't make any sense," says John Logsdon, professor emeritus at George Washington University's Space Policy Institute. "When Kennedy said he'd get to the moon by the end of the decade, he actually meant 1967, and he thought he'd still be president."

Kennedy wasn't hamstrung by budget issues, however, or by a hostile Congress, and Obama's space team is quick to

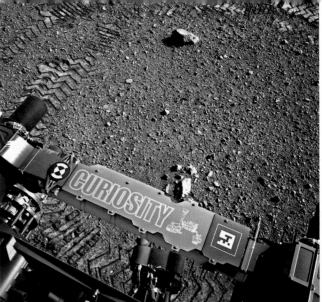

7 Aug 12
Mars isn't just a beige wasteland! There's also taupe and tan and kind of yellowish brown!! It's the most colorful planet!! SCIENCE!!

—From a sendup of Curiosity's Twitter feed by its tongue-in-cheek twin, @SarcasticRover

point that out. "I would be thrilled if we could land on Mars in the 2030s, and I truly believe that is within the capability of this country," says John Grunsfeld, head of the NASA science directorate and a five-time shuttle astronaut. "I don't believe it is necessarily within the capability of this country with a flat budget."

For the unmanned program, a flat budget would actually be an improvement. Funding for Mars missions is set to fall from $587 million in 2012 to $360.8 million in 2013. A new Mars orbiter will launch in 2013, but after that, no missions are scheduled until 2018 and 2020—maybe. Says Grunsfeld: "We can just barely afford those missions."

What any nation can afford, of course, is at least partly a function of what it chooses to afford, even in straitened circumstances. The genius of Kennedy's commitment to a lunar landing before 1970 was its simplicity—a single goal and a specific deadline. The current plan, with ever-changing destinations and dates, has none of that New Frontier clarity. And NASA's budget is a piddling 2 percent of what the Pentagon gets.

But the extraordinary success of Curiosity masks a far greater truth about space exploration: it takes monomaniacal commitment and a high tolerance for failure. In their own way, the JPL peanuts are a reminder of that fact. It's unimaginable in today's attention-deficit political climate that the agency would ever have launched Ranger 7 after the failures of Rangers 1 through 6. But NASA had to master unmanned crash landings before it could master unmanned soft landings. And it had to master unmanned soft landings before Neil Armstrong could set his boot onto the Sea of Tranquility, a mere five years after Ranger 7 made its suicide plunge into the Sea of Clouds. The exploration of Mars has followed a similar pattern. Some Mars probes—the Viking landers, for example—have been smashing successes. Others, such as the Mars Climate Orbiter, have failed ignominiously. That's the way science progresses: incrementally, patiently, and, ultimately, spectacularly. Some of America's grandest moments have come when we've trusted that fact.

GOTCHA! *Jupiter scoops up a lot of asteroids.*

When Jupiter Takes a Hit, Earth Can Get Off Easy

In the early hours of September 10, 2012, amateur astronomers spotted a flash of light on Jupiter—almost certainly a fireball triggered by a comet or asteroid about the size of a schoolbus, vaporizing as it smashed into the giant planet. That's hardly an unusual event: Jupiter sits right next to the asteroid belt, where literally millions of rocky chunks ranging from nearly 600 miles across down to the size of small boulders whirl chaotically around the sun.

Comets whiz by Jupiter all the time as well, giant snowballs plunging toward the sun from their natural home out beyond Pluto. In 1994, in fact, Comet Shoemaker-Levy 9 broke into more than 20 fragments that peppered Jupiter one after the other. Jupiter took it all in stride. But when a good-size comet or asteroid strikes Earth, there can be hell to pay: it was just such an impact 65 million years ago that many scientists believe wiped out the dinosaurs. Smaller, less deadly objects hit the planet frequently, but another world-threatening space invader, astronomers say, is only a matter of time.

If not for Jupiter, however, it could happen a lot more often. The planet is so huge, and its gravity so powerful, that it pulls in or deflects comets and asteroids that might otherwise threaten Earth. Indeed, astronomers suspect that any alien solar system lacking a Jupiter-like planet is probably a poor place to search for life.

HUBBLE FOUND YET ANOTHER MEMBER OF PLUTO'S LITTER, A FLYSPECK OF A SATELLITE JUST 21 MILES IN DIAMETER, WHICH WAS SIMPLY NAMED

P4.

Eta Carinae Nears Its Blazing Finish

The sun will die quietly some 5 billion years from now, but things will play out much differently for the star Eta Carinae, about 7,500 light-years from Earth. Weighing at least 100 times as much as our sun, it will explode as a supernova in a blast so violent that its flash will briefly outshine the entire Milky Way.

And it could happen any day now. "We know it's close to the end of its life," astronomer Armin Rest of the Space Telescope Science Institute told TIME. "It could explode in a thousand years, or it could happen tomorrow." Indeed, back in 1843, Eta Carinae signaled its unstable nature by flaring from its status as a nondescript star to become the second-brightest star in the sky, then fading after 20 years.

Astronomers back in the day did the best they could to observe the 20-year flare, but without modern instruments they couldn't really learn much. More recent observations have given theorists a trove of new information to work with—and in the next few years, said Rest, "we'll be getting more observations, and they'll keep getting better."

If Eta Carinae is going to blow imminently, the obvious question for civilians is whether Earth is in mortal danger. Fortunately, the answer is no. At 7,500 light-years, the intense radiation from even a powerful supernova would lose its punch by the time it reached us. All we'll experience is the most spectacular light show in many centuries.

Even though the theorists haven't weighed in on exactly when it will happen, Rest has reason for hope. A star in a nearby galaxy had a flare-up similar to Eta Carinae's 1843 outburst. "And then," he said, "a few years later...kaboom!"

IF THE GIANT STAR IS ABOUT TO EXPLODE, THE QUESTION FOR US IS WHETHER EARTH IS IN MORTAL DANGER. FORTUNATELY, THE ANSWER IS NO.

SOON TO BLOW *Unstable Eta Carinae could go supernova any time now.*

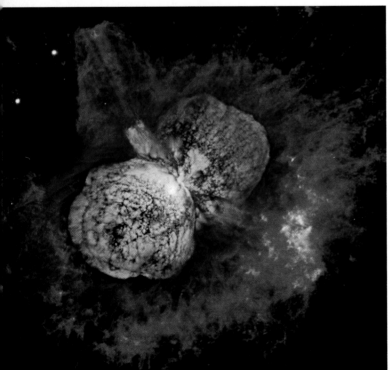

WET SPOT *False-color image of Shackleton Crater.*

A Lunar Water Faucet

The moon has no real "dark side": the face we never see gets just as much sunlight as the face that's permanently turned in our direction. But the moon has dark places—low-lying spots, often inside craters, mostly at the poles—that are permanently shaded by the surrounding terrain. If there happened to be patches of ice in these perpetually shady spots, they could survive indefinitely.

Sure enough, NASA's Lunar Reconnaissance Orbiter has found that the dark places at the lunar poles have up to 22 percent of their surfaces covered with ice. It probably rode in on comets, in the same kind of bombardment that many astronomers believe helped fill Earth's oceans.

Our planet's balmy temperatures allowed the ice to become water, and our comparatively powerful gravity kept the water in place. The small, airless moon would have frittered most of its ice right back into space—except at the poles.

That's good news for future astronauts. A permanent lunar base would need a steady supply of water, but trucking it over from Earth would be prohibitively expensive. Knowing that there would be a steady supply on hand for drinking, raising food in greenhouses, and even making rocket fuel, on the other hand, would allow space planners to check off at least one essential box long before we even consider a lunar outpost. Now all we need is the will, the wallet, and the technical know-how to check off all the rest.

For Voyager 1, the Solar System May Soon Be History

For all the astonishing feats achieved by interplanetary robotic explorers—landings on Venus, Mars, and Saturn's moon Titan; orbital reconnaissance of all the major planets and dozens of moons; and mission en route to Pluto—no spacecraft has ever left the solar system entirely.

But it may happen soon. Voyager 1, launched way back in 1977 and now nearly 11 billion miles from home, is about to become the first human creation ever to cross the invisible boundary separating our solar system from interstellar space. A surge of new data shows that the probe has seen a huge uptick in galactic cosmic rays—a key sign that Voyager is indeed about to leave the building. "The laws of physics say that someday Voyager will become the first human-made object to enter interstellar space, but we still do not know exactly when that someday will be," said Ed Stone, a Caltech physicist and Voyager project scientist, in a NASA statement.

Even at its current speed of 38,000 miles per hour, it would take Voyager 80,000 years more to reach the nearest stars, in the Alpha Centauri system, even if it were aimed in their direction. As for the golden phonograph record Voyager carries, bearing sounds and images explaining who built the probe and what we're really like, it's impossible to guess how many eons it might be before a spacefaring alien might pick up this cosmic message in a bottle, let alone whether he, or she, or it, might be able to read it. But leaving the solar system is a major milestone nonetheless—humanity's first, tentative step into the space between the stars.

INTREPID TRAVELER

Voyager 1 completed its main mission—close-up encounters with Jupiter and Saturn—back in 1980. But 32 years later the tough little probe, still going strong, is poised to achieve another milestone: becoming the first spacecraft to leave our solar system entirely and enter the unexplored realm of interstellar space.

VOYAGER 1
WAS ORIGINALLY
LAUNCHED IN

1977.

TRAVELING AT A SPEED
OF ABOUT 38,000
MILES PER HOUR,
IT IS ESTIMATED TO BE
ALMOST 11 BILLION MILES
AWAY FROM EARTH.

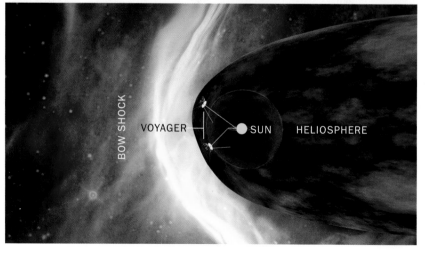

BOW SHOCK   VOYAGER   SUN   HELIOSPHERE

SUPERLONG EXPOSURE *A view of galaxies so distant they've never been seen.*

The Search for Faraway Moons

From the moment astronomers began finding planets around distant stars in the mid-1990s, they began talking about moons that might orbit those alien worlds. Now, thanks to the planet-hunting Kepler space telescope, the search for so-called exomoons is officially on.

Kepler looks for the dimming caused as a planet passes in front of its star—what astronomers call a transit. If the planet has a moon, the dimming will change subtly from one orbit to another, thanks to the complex gravitational dance of planet, star, and moon. "By combining all of this," says Harvard astrophysicist David Kipping, who leads the effort, "you can measure the mass of all three."

Once you know the mass of the planet and the moon, in particular, you can determine their density to figure out whether they're made mostly of gas, ice, or—most tantalizing for those searching for alien life—rock, just like Earth. "I've been working on the theory for a long time," says Kipping. "The last few months have been the most exhilarating time of my career." The next few should be even better.

THANKS TO THE ORBITING KEPLER SPACE TELESCOPE, THE PROSPECT OF

FINDING EXOMOONS

HAS FINALLY COME WITHIN REACH.

New Arrival: A Baby Picture of the Universe

The frustrating rule of thumb in astronomy is that the farthest objects are also the dimmest, which means it's really hard to spot the most mysterious and intriguing things in the cosmos. Now the Hubble telescope has managed to do just that: returning to a tiny patch of sky repeatedly over a 10-year campaign—a time exposure adding up to more than 500 hours of observation—the venerable Hubble has given astronomers an unprecedented look at more than 13 billion years of cosmic history, all in one image.

Known as the eXtreme Deep Field, or XDF, the picture offers up thousands of galaxies billions of light-years away, stretching back almost to the time when the first stars began to shine—because it's taken that long for the light from these early galaxies to reach us. "The XDF is the deepest image of the sky ever obtained and reveals the faintest and most distant galaxies ever seen," astronomer Garth Illingworth of the University of California, Santa Cruz, and the principal investigator on the project, said in a NASA statement.

As for what it means...well, that's not at all clear yet. Powerful as it is, the Hubble is acting as a mere spotting scope for the much bigger ground-based telescopes that will now gaze deeply at these newly discovered—but very old—targets, teasing out the secrets of what the universe was like when the very first stars began to shine in the blackness of space.

THEY'RE OUT THERE SOMEWHERE *Since alien planets are common, moons that orbit them must be too.*

NEARBY WORLD *A telescope in Chile made the discovery.*

THE NEW PLANET, DESIGNATED ALPHA CENTAURI B B, IS ABOUT THE SAME SIZE, AND QUITE POSSIBLY THE SAME COMPOSITION, AS EARTH.

A New Planetary 'Neighbor'

Most of the planets found orbiting stars beyond the sun—some 800 at last count, with another 2,300 or so "planet candidates," identified but not confirmed—are tens or even hundreds of light-years from Earth. Not so a brand-new world announced with great fanfare in the journal *Nature,* however. The planet, discovered by a team of Switzerland-based astronomers, is orbiting happily in the Alpha Centauri system, whose three stars, a mere four light-years away, are the very nearest to our own solar system.

Four light-years is still pretty far (it works out to more than 20 trillion miles), but if you want to study alien planets for signs of life, the closer the better—and, says planet hunter Greg Laughlin of the University of California, Santa Cruz, "the odds against having such a good target so close by are a thousand to one." Even better, the new planet, known as Alpha Centauri B b (the first "B" identifies the star, the second the planet), is about the size of Earth—far smaller and more homey than most exoplanets discovered to date. The bad news: it's also so close to its star that the surface could well be molten lava.

But planet hunters have learned that where there's one planet, there could be more, and one of them could be life-friendly. A star system that has long intrigued sci-fi writers now has scientists taking notice.

On Mercury, More Action Than We Thought

Mercury is the solar system's Greenland. We know it's there, we've heard some interesting stuff about it, but we just don't give it a whole lot of thought—and "we" long included NASA.

But that changed in 2004, when NASA launched the Messenger spacecraft toward the solar system's innermost planet. The probe went into orbit around Mercury in 2011 and has been surveying it with laser altimeters, gravitational instruments, cameras, and more ever since. And it turns out that the tiny planet is a far more dynamic—and far less predictable—place than we once believed.

It turns out, for example, that where Earth has a thick, rocky mantle surrounding a compact core of iron, Mercury is little more than a rocky rind coating a mostly metallic sphere. That rind should have solidified very quickly. But Messenger discovered some relatively blemish-free patches on a planetary surface that scientists expected would be uniformly peppered with craters. That means that at least some lava remained molten long enough to repave at least part of the surface well after all of Mercury's rock should by rights have cooled.

Mercury's surprising geology will never quite match Saturn's rings or Jupiter's watery moons or the rivers and oceans that once graced Mars. But there's something wonderfully plucky about this small and sun-blasted world that simply refuses to settle down.

RELIEF MAP *Red means high altitude, purple low.*

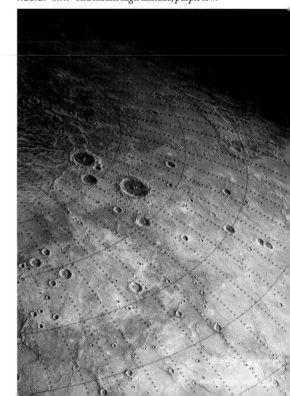

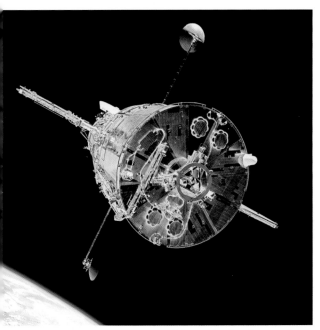

SPARE PARTS *Two new Hubbles could replace the original.*

Astronomers Get a Windfall

The Hubble Space Telescope is still going strong, but at 22 years old it's beginning to wear out. With no shuttles to refurbish it, this aging technological marvel will eventually die. That's why astronomers were so excited to learn in the spring of 2012 that two pristine Hubble-class telescopes have been sitting, unsuspected, in a warehouse in Rochester, New York. Originally built for the National Reconnaissance Office, they were designed to look down on Earth, not out into space, but the spy agency has moved on to even better technology—and now scientists could reap the benefit.

It's not clear what astronomers will do with these astonishing gifts—and indeed, there's only a general consensus on how they might use just one of the telescopes, never mind both. "A few of us began discussing this quietly when we first learned about it," says Princeton's David Spergel, "and now we'll be talking to the wider community about how to use them." It will take a while, he says, before there's a concrete plan on how to move forward. Until that happens, astronomers will be basking in improbable good luck that they've been given two shiny, brand-new toys to play with.

THE $1.5 BILLION HUBBLE
IS EXPECTED TO GO OUT OF SERVICE SOMETIME AROUND 2014, 24 YEARS AFTER ITS LAUNCH.

That Supermoon Wasn't So Super

The "supermoon" in May 2012 and the Mayan apocalypse scheduled for seven months later had one big difference and one big similarity. The difference: the supermoon was real, while the apocalypse was a fevered fantasy. What the two shared, however, was an enormous dose of hype.

May's event was unusual simply because the moon's closest approach to Earth in its slightly oval orbit coincided with the timing of the full moon. When that happens, it looms a bit bigger in the sky than usual.

Only a bit, though. It meant, tweeted Neil deGrasse Tyson, director of the Hayden Planetarium, that "the impending Supermoon is to an average full Moon what a 16" Pizza is to a 15" Pizza. So chillax."

Without the hype, which included the truly nonsensical claim that the supermoon would not only dazzle the eye but perhaps trigger earthquakes as well, nobody might even have noticed that anything was different. Yes, the moon looked surprisingly huge if you caught it just after it rose. But it nearly always does, thanks to the well-known "moon illusion": the moon tends to look bigger when it's close to the horizon, where you can compare it at a glance with buildings or trees or mountains. Once it rises higher into the sky where it's surrounded by nothing but a lot of blackness, it seems to shrink. It's always the same size, of course.

Nevertheless, even while he ranted about the silly publicity, the popular astronomy blogger Phil Plait urged people to go out and take a peek, not because the moon was especially super but "because it's pretty, and it's ours, and it's always worth a look." Few who took his advice regretted it.

ILLUSION *It looked huge at first, before its ascent into the sky.*

The Mind

FOOTBALL'S CONCUSSION FACTOR ▪ OBESITY AND THE BRAIN ▪ INFANT ANESTHESIA
WHY SEX DOESN'T GROSS YOU OUT ▪ IS SELFISHNESS INNATE? ▪ MILITARY SUICIDES
HOW TO DIAGNOSE ALZHEIMER'S ▪ SHOPPING WHILE HUNGRY ▪ MEASURING AUTISM

The Upside
of Being an Introvert

(And why extroverts are overrated.)

By Bryan Walsh

I'm in the bathroom of the American Embassy in Tokyo, and I can't leave. Somewhere in the elegant rooms beyond, the ambassador is holding his annual holiday party. Diplomats from around the world, U.S. military personnel, and reporters are mingling, sipping champagne, and picking at hors d'oeuvres. As TIME's Tokyo bureau chief, I should be there, trolling for gossip or mining potential sources.

And for 20 minutes or so after arriving, despite the usual nerves, I did just that. But small talk with stiff-backed strangers at a swanky cocktail party is by far my least favorite part of my job. Send me to a famine or a flood, and I'm comfortable. A few rounds of the room at a social event, however, leave me exhausted. So now and then I retreat into the solitude of the bathroom, watching the minutes tick by until I've recovered enough to go back out there.

My name is Bryan, and I'm an introvert. If this scene sounds familiar to you, then chances are that you're one too.

We're not alone, even if it sometimes feels that way. By some estimates, 30 percent of all people fall on the introvert end of the temperament spectrum—but it takes some explaining to understand just what that label means. For one thing, introverted does not have to mean shy, though there is overlap. Shyness is a form of anxiety characterized by inhibited behavior. It also implies a fear of social judgment that can be crippling. Shy people actively seek to avoid social situations, even ones they might want to take part in, because they may be inhibited by fear. Introverts shun social situations because, Greta Garbo style, they simply want to be alone.

"Introverted people aren't bothered by social situations," says Louis Schmidt, director of the Child Emotion Laboratory at McMaster University in Ontario. "They just prefer not to engage." While extroverts draw energy from mingling with large groups of people—picture former president and extrovert in chief Bill Clinton joyously working a rope line—introverts find such social interactions taxing.

Just being an introvert can also feel taxing—especially in America, land of the loud and home of the talkative. From classrooms built around group learning to open-plan offices that encourage endless meetings, it sometimes seems that the quality of your work has less value than the volume of your voice.

But all that discounts the hidden benefits of the introverted temperament—for workplaces, personal relationships, and society as a whole. Introverts may be able to fit all their friends in a phone booth, but those relationships tend to be deep and rewarding. Introverts are more cautious and deliberate than extroverts, but that means they tend to think things through more thoroughly, which means they can often make smarter decisions. Introverts are better at listening—which, after all, is easier to do if you're not talking—and that in turn can make them better business leaders, especially if their employees feel empowered to act on their own initiative. And simply by virtue of their ability to sit still and focus, introverts find it easier to spend long periods in solitary work, which turns out to be the best way to come up with a fresh idea or master a skill.

Introversion and extroversion aren't fixed categories—there's a personality spectrum, and many people, known as ambiverts, fall in the gap between the two traits—but they are vital to our personalities. "There's a subtle bias against introverts, and it's generating a waste of talent and energy and happiness," says Susan Cain, author of *Quiet: The Power of Introverts in a World That Can't Stop Talking.* It may be time for America to learn the forgotten rewards of sitting down and shutting up.

Scientists have begun to learn that the introverted or extroverted temperament seems strongly inborn and inherited, influencing our behavior from not long after we're out of the womb.

That was the conclusion of a pioneering series of experiments by Harvard developmental psychologist Jerome Kagan. In a 1989 study, he and his colleagues gathered a sample group of 500 four-month-old infants and exposed them to new experiences in the lab, including popping balloons, colorful mobiles, and the smell of alcohol on cotton swabs. About 20 percent of the infants reacted intensely to the stimuli, crying and pumping their arms. About 40 percent stayed relatively quiet, and the remaining 40 percent fell between the two extremes.

Kagan's theory about their future development was contrary to what you might think. He predicted that the infants who had the most noticeable responses—the group he called high-reactive—would likely be introverted as adolescents, whereas low-reactives would likely be extroverted. When he brought his subjects back into the lab as they grew older, his hypothesis proved true: high-reactive infants matured into more inhibited, introverted teenagers. People who are introverts by nature, it turns out, may simply have a lower threshold for stimulation than others. It doesn't take too many popped balloons and crowded rooms before they learn to compensate by keeping a low, quiet profile, conserving their limited energy. Meanwhile, extroverts are a little bit like addicts who are always in search of a high, seeking out stimuli—in the healthier form of social situations—that would make an introvert's head ring.

Caution, inhibition, and even fearfulness may be healthy—and smart—adaptations for the overstimulated person, but they're still not characteristics many parents would want in their children, especially in a society that lionizes the bold. So it's common for moms and dads of introverted offspring to press their kids to be more outgoing, lest they end up overlooked in class and later in life.

QUIET (AND LOUD) GIANTS: IT TAKES BOTH KINDS TO MAKE HISTORY

Introverts

MOHANDAS GANDHI
Revolutionary
Gandhi changed the direction of an entire nation, but he was always an inward-looking person, at ease on his own—and in his own skin.

JOE DIMAGGIO
Baseball Hall of Famer
The Yankee great was not shy at the plate, but he withered in the glare of his marriage to Marilyn Monroe and was reticent in retirement.

HILLARY CLINTON
Secretary of state
The debate in 2008 was whether Clinton was likable. If that quality came hard for her, it's because she lacks the gregariousness of her husband. The very public woman is actually very private, too.

WARREN BUFFETT
Magnate
Buffett has said that one of the most important traits for investing is having the right temperament—and his introverted, cautious personality clearly works.

MANMOHAN SINGH
Prime minister
For a man who governs the world's second-most-populous country, Singh is notably self-effacing, a lifelong technocrat who isn't always comfortable in rough-and-tumble Indian politics.

BILL GATES
CEO, philanthropist
Gates has the ferocious focus that allowed him to spend thousands of hours writing code, and he still seems more comfortable with technology than he is with people.

MOTHER TERESA
Nun, missionary
People were often shocked at just how soft-spoken this nun from Albania could be in person, but that didn't stop her from becoming a larger-than-life figure.

That, however, can be a mistake—and not just because our temperaments are difficult to change fundamentally.

The very fact that introverts are more sensitive to their environment often means they're fully aware that they appear out of step with the expectations of others, and they can easily internalize that criticism. But introverts also have tremendous advantages. Sure, there are thrills to be found in the situations that extroverts crave, but there are dangers too. Extroverts are more likely than introverts to be hospitalized as the result of an injury, for example, and they're more likely to have affairs or change relationships frequently, with all the collateral damage that can entail.

The introvert advantage isn't only about avoiding trouble. Florida State University psychologist K. Anders Ericsson believes that deliberate practice—training conducted in solitude, with no partner or teammate—is key to achieving transcendent skill, whether in a sport, in a vocation, or with a musical instrument. In one study, Ericsson and some of his colleagues asked professors at the Music Academy in Berlin to divide violinists into three groups, ranging from those who would likely go on to professional careers to those who would become teachers instead of performers. The researchers asked the violinists to keep diaries and found that all three groups spent about the same amount of time—more than 50 hours a week—on musical activities. But the two groups whose skill levels made them likelier to play well enough to perform publicly spent most of their time practicing in solitude.

The trouble is, fewer and fewer of us have time for solitary contemplation and practice anymore. It's not just the assault of email, cell phones, and social media. The very geography of the American workplace is designed to force people together. Some 70 percent of American workers spend their days in open-plan offices, with little or no separation from colleagues; since 1970 the average amount of space allotted to each employee has shrunk from 500 square feet to 200 square feet. Much of this is done in the name of collaboration, but enforced teamwork can stifle creativity.

It's not just introverts who suffer when work becomes an endless series of meetings and brainstorming sessions. Anyone who has spent time in any organization knows that there is rarely a correlation between the quality of an idea and the volume at which it is presented. Defying the loudest speaker—and the groupthink that tends to build around that person—can be painful for anyone. When people oppose group consensus, for example, the amygdala regions of their brains tend to become stimulated, signaling fear of rejection.

To break that pattern requires the right kind of leader, and the right kind of leader may be an introverted one. By one estimate, 40 percent of high-powered American businesspeople fall on the introvert end of the spectrum, a group that appears to include the likes of Bill Gates, Charles Schwab, and Google CEO Larry Page. The ability to assess risk and remain focused on the long term can pay off big in the boardroom. So can the capacity for listening, a trait that can be too easily lost at the exalted level of the C suite. "Introverted leaders tend to be more detail-oriented and better able to hear their employees," says Jennifer Kahnweiler, an executive coach and author of *The Introverted Leader: Building on Your Quiet Strength.*

But just because all of us have powerful inborn traits doesn't mean we can't stretch the limits of our personalities when the stakes are high enough. From the moment introverts wake up to the second we go to sleep—preferably after relaxing with a book in bed—we live in an extrovert's world, and there are days when we'd prefer to do nothing more than stay at home. But while our temperaments may define us, that doesn't mean we're controlled by them—if we can find something or someone that motivates us to push beyond the boundaries of our nerves. I'm happy to be an introvert, but that's not all I am. –*With reporting by Cleo Brock-Abraham*

Extroverts

BILL CLINTON
President
The White House has mostly been occupied by extroverted men, perhaps none so much as Clinton. Crowds that would exhaust others serve only to energize him.

MARGARET THATCHER
Prime minister
The Iron Lady was no shrinking violet. Thatcher's extroverted personality helped her bull her way through British sexism to become the country's first female P.M.

STEVE JOBS
Innovator
Jobs was a born salesman, comfortable with demanding the world's attention. An exacting boss, he knew how to get the best ideas from his workers, even if he earned the enmity of some.

BORIS YELTSIN
President
Yeltsin had a flair for the dramatic—like standing on a Soviet tank as he defied a 1991 coup—and loved the stage. But his injudicious impulsiveness ruined his time in office.

MARIE ANTOINETTE
Queen
The original royal party girl did not endear herself to her French subjects, nor was she a great fit for the painfully shy Louis XVI.

MUHAMMAD ALI
Boxer
He called himself "The Greatest," he was as much a performer as a slugger, and it took Parkinson's disease to slow him down—a little. Sound like an introvert to you?

WINSTON CHURCHILL
Politician
Churchill had a limitless supply of energy—and a limitless taste for drink. But he didn't let that get in the way of his work, including writing—he won the Nobel Prize in literature for his memoirs.

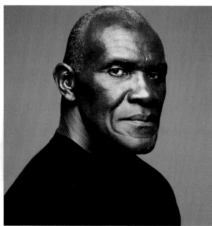

KYLE TURLEY, TED JOHNSON, AND HARRY CARSON *(Clockwise from top left) Retired NFL players who experienced head injuries are speaking up about the long-term effects of playing-day concussions on their health. Many former players have sued the NFL, charging the league with failing to inform them of the dangers of blows to the head.*

PRO FOOTBALL PLAYERS ARE

4 TIMES

AS LIKELY AS NONPLAYERS TO DIE OF A NEURODEGENERATIVE BRAIN DISEASE SUCH AS ALZHEIMER'S OR ALS.

Football's Dangers to the Brain

Years of pounding on the gridiron takes its toll, not just on the body but on the brain as well. After analyzing death certificates of 3,439 former National Football League players who participated in at least five seasons between 1959 and 1988, scientists found that players are three times as likely as others to die of a neurodegenerative brain disease, and four times as likely to succumb to Alzheimer's or amyotrophic lateral sclerosis (ALS), also known as Lou Gehrig's disease. And as alarming as the data are, they may underestimate the dangers to the brain, since the game now features bigger, stronger, and faster players than those who took to the field decades ago.

By 2007 about 10 percent, or 334, of the participants had died. Overall, the study, published in the journal *Neurology,* shows that players appeared to be in exceptional health: at any given time they were half as likely to die as other men their age. But the physical demands of professional football make them more vulnerable to concussions, and data show that those who suffer concussions are more likely to develop mild cognitive impairment, memory problems, or chronic traumatic encephalopathy (CTE), a degenerative disease that causes dementia and looks a lot like Alzheimer's.

The authors note that while they looked for Alzheimer's, Parkinson's, and ALS as causes of death, it's possible that some of the players in their study could really have died from CTE. But because CTE is not yet listed as an official cause of death in the International Classification of Diseases, it wouldn't appear on death certificates. Instead, many of the players may have been listed as dying of Alzheimer's or ALS.

In response to growing concerns about the game's dangers—the NFL currently faces about 140 lawsuits from about 3,400 former players or their families who say the league downplayed and misrepresented the risks of head injury and concussion to players (which the NFL denies)—the league pledged $30 million for medical research to the Foundation for the National Institutes of Health. The money will be overseen by the NIH and may go toward research on CTE and concussion management and treatment.

Body Weight and Cognitive Decline

Obesity is already implicated in a number of health problems, from heart disease to diabetes. But now researchers are adding another ailment to the list: cognitive decline.

In a study of 6,401 people ranging from normal weight to obese, researchers report in the journal *Neurology* that there is a significant connection between body weight, along with some of its accompanying metabolic changes, and the brain. Thirty-one percent of the participants, whose average age was 50, had two or more metabolic abnormalities—including high blood pressure, low levels of HDL, high blood sugar, and high triglycerides.

At three points over the 10-year study, the participants took memory and cognitive tests in which they were asked to write down, for example, all the words they could think of starting with the letter S, in one minute. Comparing these results over time, the researchers found that those who were obese and metabolically abnormal registered a 22.5 percent faster cognitive decline than normal-weight participants with no metabolic abnormalities. Obese participants who were metabolically normal also saw faster declines in their cognitive skills compared with metabolically normal adults with a healthy weight.

The finding contradicts the emerging idea that people can be obese and, as long as they're metabolically normal, also be healthy. The changes in the body that occur with weight gain may not necessarily show up as a metabolic condition but could be damaging nonetheless. It could be that inflammation is involved somehow. And it's possible that other factors typically associated with obesity, such as lack of exercise and smoking, are also affecting the brain.

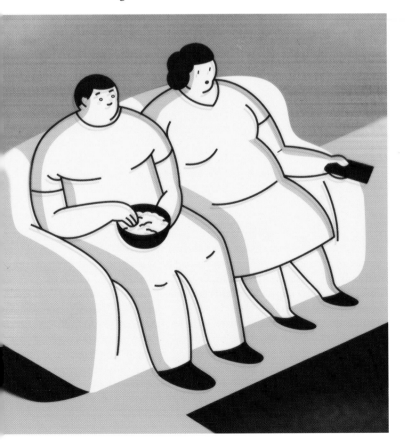

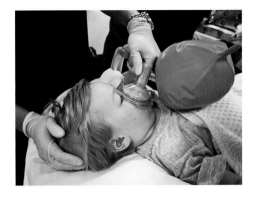

Does Infant Anesthesia Harm Children Later?

Millions of infants go under the knife every year to treat or avoid ailments, some of which could become life-threatening, but those procedures may come with a high price. Increasingly, research links exposure to anesthesia in early childhood to long-term effects on kids' brain development. In the latest analysis, researchers report in *Pediatrics* that children who had received anesthesia before age 3 were 87 percent more likely to show language disabilities and nearly 70 percent more likely to have cognitive problems at age 10, compared with those who had not been exposed to anesthetic drugs. Even a single exposure to anesthesia was associated with increased risk. Hardest hit are language and abstract-reasoning skills, says researcher Caleb Ing, an anesthesiologist at Columbia University, while behavior and motor skills remain relatively unaffected.

The findings don't prove that anesthesia directly causes problems in brain development, but since most of the children in the study received anesthesia for benign procedures like tonsillectomies, it's unlikely that the deficits are due to the underlying medical conditions that required surgery in the first place. So for parents who are worried about their children's long-term health, it's worth asking their doctor about the necessity of the operation. In many cases, the benefits will outweigh the still-unproven risks of anesthesia.

CHILDREN WHO RECEIVED ANESTHESIA EARLY IN LIFE WERE

87%

MORE LIKELY TO SHOW

LANGUAGE DISABILITIES.

The Arousal Factor: Why Sex Doesn't Gross You Out

Think about it. Sex is actually sort of disgusting, what with all the sweat, saliva, fluids, and smells, isn't it? So much so that a group of researchers from the Netherlands got to thinking: how do people enjoy sex at all?

According to their small new study (of 90 women), people—at least females—may be able to get over the "ick" factor associated with sex by getting turned on. Sexual arousal overrides the disgust response, and even allows women to engage in behaviors they might normally find repugnant.

For the study, intercourse in the lab wasn't involved, but instead women were divided into groups that watched one of three videos: an erotic one, a stimulating but nonsexual one (of skydiving, for example), or a neutral clip of a train. Then they were asked to perform a series of unpleasant tasks, including drinking from a cup with a bug in it (the bug was fake).

The women who watched the erotic video rated the unpleasant tasks as less disgusting than their counterparts did, and were also more likely to complete a larger number of them, suggesting that sexual arousal not only dampens the disgust response but also influences how much women are willing to do.

That supports the idea that sexual arousal lowers inhibitions and could mean that some sexual dysfunction, particularly among women, may be treated by boosting arousal. Foreplay, anyone?

Are We Wired to Be Selfish or Altruistic?

Under the survival-of-the-fittest mentality, man evolved to compete for precious resources, making us a relatively selfish bunch (yielding a potential mate to a rival, for example, doesn't do your genes any good in seeing another generation). But those theories fail to account for the fact that humans could not have survived without the occasional charity and social reciprocity of the group, too.

And studies show that this altruism is innate. Even 18-month-old toddlers will almost always try to help an adult who is visibly struggling with a task, without being asked to do so: if the adult is reaching for something, the toddler will try to hand it to him. Another study found that 3- to 5-year-olds tend to give a greater share of a reward (stickers, in this case) to a partner who has done more work on a task—again, without being asked—even if it means they get to keep less for themselves.

How is all this helping hard-wired? Our stress systems themselves seem to be designed to connect us to others. They calm down when we are feeling close to people we care about—whether related to us or not—and spike during isolation and loneliness. Even short periods of solitary confinement can derange the mind and damage the body because of the stress they create.

Of course, that doesn't mean humans are never selfish, but those bouts of self-centeredness seem to be balanced with a good dose of the Good Samaritan as well.

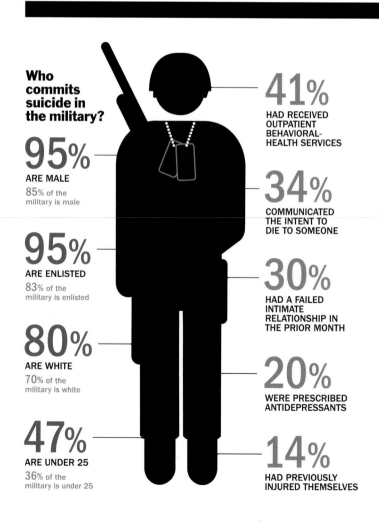

Who commits suicide in the military?

95%
ARE MALE
85% of the military is male

95%
ARE ENLISTED
83% of the military is enlisted

80%
ARE WHITE
70% of the military is white

47%
ARE UNDER 25
36% of the military is under 25

41%
HAD RECEIVED OUTPATIENT BEHAVIORAL-HEALTH SERVICES

34%
COMMUNICATED THE INTENT TO DIE TO SOMEONE

30%
HAD A FAILED INTIMATE RELATIONSHIP IN THE PRIOR MONTH

20%
WERE PRESCRIBED ANTIDEPRESSANTS

14%
HAD PREVIOUSLY INJURED THEMSELVES

The Military's Toughest War

One a day. Among those serving on active duty in the U.S. military, one will take his life (and almost all are male) every day. Among all veterans a suicide occurs every 80 minutes, around the clock.

These are sobering facts, ones that stymie the military's best efforts to understand why. More U.S. military personnel have died by suicide since the war in Afghanistan began than have died fighting there. While it's hard to come by historical data on military suicides—the Army has been keeping suicide statistics only since the early 1980s—there's no denying that the current numbers constitute a crisis.

No program, outreach, or initiative has worked against the surge in Army suicides, and no one knows *why* nothing works. The Pentagon allocates about $2 billion—nearly 4 percent of its $53 billion annual medical bill—to mental health. But that's clearly not enough. The disturbing trend defies the most logical causes; while those who deployed to combat zones like Iraq or Afghanistan might be considered at highest risk of the type of mental trauma that might trigger suicide, nearly a third of the suicides from 2005 to 2010 were among troops who had never deployed.

Why? One theory of suicide holds that people who feel useful, who feel as if they belong and serve a larger cause, are less likely to kill themselves. That would explain why active-duty troops historically had lower suicide rates than civilians. But now experts who study the patterns wonder whether prolonged service during wartime may work

against soldiers when their tours are over. Service members who have bonded with their units can have trouble once they are at a post back home, away from the routines and rituals that arise in a close-knit company.

For war veterans, it's not war that turns out to be hard; it's peace. In addition to missing the camaraderie and connection of being part of a platoon or unit, many feel unhinged by the lack of routine in civilian life. Whatever the cause, the issues may be exacerbated by the pride and protocol of a warrior culture in which needing help, particularly for a mental illness, is seen as a sign of weakness. Widows of military personnel who have taken their own lives tell similar stories—of the challenge not only in convincing their spouses to seek help but in finding a receptive military that could have provided the therapy their loved ones so desperately needed.

The Army says there is no way to tell how well the suicide-prevention programs it does have are working, but it estimates that without such interventions the number of suicides could have been four times as high. Resources are a perennial problem: thinner ranks of mental-health professionals nationwide mean the military is getting by with fewer properly trained people to handle the delicate issues of the mind. That's little consolation to Rebecca Morrison, whose husband, an Apache helicopter pilot, shot himself while attempting to get therapy for depression. "My husband did not want to die," she says. "Ian tried to get help—six times in all. He just wanted to be healthy."

Grocery Shopping While Hungry

You shouldn't go grocery shopping when you're hungry, because you're more likely to buy foods loaded with calories and sugar. But do you know why?

A study led by Yale University and University of Southern California researchers may help. They hooked up 14 men and women, ranging from normal weight to obese, to a device that controlled their glucose levels intravenously, in order to simulate hunger and fullness, and then sent them through a functional MRI (fMRI) scanner that recorded their brain activity. While in the machine, the participants saw pictures of high- and low-calorie foods, as well as nonfood objects.

When the researchers kept the participants' glucose levels near normal, the regions of the brain in the prefrontal cortex that control emotions and impulses, like craving high-calorie junk food, were more active. But when glucose levels were dropped, deeper areas such as the hypothalamus, thalamus, and nucleus accumbens, which govern motivation, reward, and addiction, lit up instead. In other words, our bodies are wired to crave dense foods more when our stored energy resources are depleted.

Interestingly, in the obese participants the signal to crave high-calorie eats became the status quo, which could explain why the reward system continues to signal their brains to eat, even when they're full.

That may prove an important clue to understanding obesity. And if these findings are confirmed, they could lead to more effective ways to curb weight gain, beginning not on the plate but in the brain.

A STUDY SHOWS THERE MAY BE A
BIOLOGICAL DRIVER
THAT PUSHES OBESE PEOPLE TO CONTINUE EATING EVEN WHEN THEY'RE FULL.

How to Diagnose Alzheimer's Early On

In the United States, 5.4 million people are living with Alzheimer's disease. But despite the large number of patients, doctors are surprisingly ill equipped to diagnose, much less launch a treatment strategy for, those affected. Now, for the first time in nearly three decades, even non-expert physicians have a new set of guidelines to better diagnose Alzheimer's disease in the clinic. The advice includes criteria for identifying the earliest signs of the degenerative condition, even before symptoms of memory loss begin, with the hope that doctors can help patients to prepare for, and eventually treat, the disease.

Currently, Alzheimer's can be definitively diagnosed only at autopsy, when pathologists can confirm the presence of protein plaques and tangles in the brain. But the latest research suggests that the disease develops over years or even decades before the first cognitive deficits are noticeable, so the idea is to diagnose patients as early as possible. That's why the new guidelines tease apart three stages of the disease: preclinical, pre-dementia, and dementia due to Alzheimer's. While many procedures recommended in the guidelines—such as blood tests to pick up the protein plaques or imaging studies to see them—are still experimental, doctors believe the information they provide will be critical for developing more reliable ways to diagnose and treat Alzheimer's in the future.

Identifying Autism 33 Ways

Autism is most commonly diagnosed by changes in a young child's behavior: he stops talking or making eye contact, or starts to ignore those around him. But researchers are beginning to create a picture of what the autistic brain looks like to help doctors more accurately diagnose the developmental disorder.

Scientists at Boston Children's Hospital used electroencephalography (EEG) to measure brain activity in children and found 33 specific patterns that could differentiate an autistic child from a neurologically typical peer. Those with autism had activity patterns that consistently showed reduced connectivity between brain regions, especially in areas associated with language. "What is surprising is that this pattern of 33 factors was so useful from age 2 to 12," says study author Dr. Frank Duffy of the hospital's department of psychiatry.

In order for the brain to work properly, its different regions need to function together through nerves that connect them in a harmonious network. When those connections fail, problems in language or social skills can appear.

The researchers hope that the EEG, a widely used and affordable test, could eventually be deployed to identify children on the autism spectrum, particularly in developing countries or underserved areas in the United States where specialists are in short supply and autism screening is less commonplace. In addition, the imaging profiles could distinguish between the different conditions along the autism spectrum, teasing apart high-functioning children with Asperger's syndrome, for example, who may not require as much behavioral intervention as autistic children.

IT JUST AIN'T SO ...

Fish Oil's Deflated Reputation

Omega-3 fatty acids are widely touted for their ability to preserve cognitive function and memory, but a new review by the Cochrane Library finds that those effects may be overstated. Healthy elderly people taking omega-3 supplements, the new analysis found, did no better on tests of thinking and verbal skills than those taking placebos.

Previous studies have asked participants to recall how much omega-3 fat they ate, primarily from oily fish like salmon, and connected these patterns with such health factors as brain volumes and levels of beta amyloid, the telltale protein that gums up brains and impairs memory in Alzheimer's patients. But in the Cochrane review, researchers included only those studies in which participants were randomly assigned to take either omega-3s or a placebo and then tracked them over time. That way they hoped to lower the chances of biased results. Overall, the data showed that participants taking extra omega-3s performed no better on standard tests of mental abilities, memory, or verbal fluency than those who took placebos.

That doesn't mean that omega-3 fats can't help the brain somehow. The study followed volunteers for only a few years, too brief a period to see signs of cognitive decline and dementia, which often take longer to develop. In addition, the data supporting the benefits of omega-3 fatty acids on the heart are much stronger—and what's good for the heart may also be good for the brain.

Technology

GOOGLE GOGGLES ▪ HONEYCOMB ARCHITECTURE ▪ NEW NASA SPACESUIT
WHEELCHAIRS ON BOARD ▪ DEEP-SEA CHALLENGE ▪ A CARDBOARD BICYCLE
SURVEILLANCE HUMMINGBIRD ▪ BABY-SOFT BANDAGE ▪ HOLOGRAPHIC MAPPING

The Revolution Will Be Printed

*Call them 3-D printers or personal fabrication tools.
A new way to make things is about to become huge.*

BY DAVID BJERKLIE

It is a cold autumn day in lower Manhattan, and a crowd is peering inside a retail shop that looks like a cross between Santa's workshop and a gleaming new computer store. In the display window stands a floor-to-ceiling contraption releasing steel marbles that travel crazily from top to bottom through a roller coaster of tunnels. There are also giant gumball machines in the store filled with tiny (and freshly made) plastic prizes, including robots, action figures, and skyscrapers. The entranced shoppers have stumbled upon the showcase of MakerBot, a manufacturer of 3-D printers. If you don't yet know what a 3-D printer is, not only would MakerBot like to tell you (and show you), the company would also love to persuade you that you need to buy one for yourself. A 3-D printer is actually a desktop manufacturing system. Commercial machines on the market can already produce model cars, cell-phone cases, sculptures, jewelry, and dollhouse furniture, as well as practical items like replacement knobs for small appliances. But the brave new future of 3-D printers may include everything from pork chops to handguns.

3-D printers take digital designs and then "print" them out, layer after layer, with special "inks" until a three-dimensional object emerges. The inks in the MakerBot, which are loaded as spools of brightly colored filament, are either ABS (acrylonitrile butadiene styrene, which is the hard plastic of which Legos are made) or the corn-based bioplastic PLA (polylactic acid). More exotic inks used in other printers include chocolate, metals like titanium, and even living cells.

MakerBot, with its unique (so far) retail store in New York City, may be poised to be an early leader in the consumer market (it had sold 15,000 printers by late 2012), but it has

PRINTED MATERIAL *A digital design is executed with special "inks" that are applied layer after layer until an object emerges.*

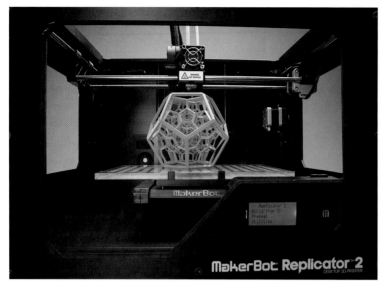

You may have heard about 3-D printers, but chances are you don't own one—yet. New and affordable models are poised to change that.

DESKTOP MANUFACTURING *MakerBot had sold 15,000 of its printers by late 2012.*

plenty of competition, in the U.S. and around the world. And while prices are steep—most models cost well over $2,000—they will surely drop. Like computers, they'll be available at a wide range of prices, from bare-bones starter kits to deluxe units with all sorts of peripherals (which the printer may be able to make itself).

Hand in glove with the new capacity to make things is a burgeoning online library of digital designs for this stuff. At MakerBot's website Thingiverse, 3-D printing enthusiasts can access and contribute to a "universe of things." But again, MakerBot is not alone. A growing number of companies, such as Shapeways, Autodesk, and Tinkercad, are offering designs, as are grab-and-go sites like Pirate Bay. As a result, the battle over copyrights and intellectual property is already brewing. (A company run by Nathan Myhrvold, the former chief technology officer of Microsoft, has been issued a patent for a system that could prevent peo-

ple from printing objects using designs they haven't paid for.)

While the prospect of manufacturing toys on demand is a kid's dream come true, the enthusiasm for 3-D printing extends to industrial giants like GE and Boeing. In the early 1900s, the power and speed of Henry Ford's assembly line revolutionized industrial production by leveraging mass-market similarity. But today, customers increasingly demand products that are specific to their needs: mass customization. By any name, 3-D printing, in the words of Michael Idelchik, vice president of advanced technologies for GE Global Research, "has the potential to fundamentally disrupt" how we make things—and transform industries in the process.

The key to mass customization is "additive" manufacturing. Michelangelo once described his technique as "I saw the angel in the marble, and carved until I set him free." That leaves a lot of marble on the floor. Similarly, modern "subtractive" manufacturing takes blocks or slabs or sheets of material and cuts, grinds, saws, or mills away everything that is not the object desired. 3-D printing, however, makes it possible to produce objects while reducing waste to essentially zero. The technology also reduces the need for inventory, since parts can be printed on demand, as well as continually improved or customized.

The techniques used by industrial giants go well be-

yond anything that MakerBot and other consumer printers are capable of. Sintering, for example, relies on a laser or electron beam to fuse successive layers of powdered metal; the technique is already being used to make parts for military and commercial aircraft, including Boeing's new 787 Dreamliners. As technology improves in terms of speed, cost, and materials, the opportunities for industrial-strength 3-D printing are expected to boom.

There is also a growing list of unexpected applications. A company called Bespoke Innovations, founded in 2009 by an industrial designer and an orthopedic surgeon, is using 3-D printing to produce customized outer coverings (called fairings) for prosthetic limbs. The company's mission is to help people "emotionally connect with their prosthetic limbs, and wear them confidently as a form of personal expression." Choc Edge, a British company, offers enticing but pricey chocolate printers. And researchers at Cornell University's Creative Machines Lab are working with chefs to explore how printers can create new culinary tastes, textures, colors, and forms as well as enhance nutritional value.

And that's just for starters. At the annual TEDMED conference for innovation in 2011, University of Missouri bioengineer Gabor Forgacs presented—and then ate—a "pork chop" that had been produced by a bioprinter equipped with pig-cell ink that had been grown in vitro. Forgacs and his son Andras hope to turn 3-D printed meat, as well as printed leather, into a business that would lessen the environmental impact and ethical objections of raising meat the old-fashioned way. (Their company, Modern Meadow, even has the blessing of Ingrid Newkirk, president of People for the Ethical Treatment of Animals.)

Bioprinting has its origins in the field of regenerative medicine (which happens to be Forgacs's day job and the focus of Cornell's lab as well). Tissue engineers have long dreamed of growing tissues in the lab that could then be used to repair or replace failing organs in the human body. Researchers are now using 3-D printing techniques to create biotubing similar to blood vessels, implant bioprinted "scaffolds" into the jaw and hip bones of rats and rabbits, and print "skin" for mice and pigs. Human trials of bioprinted tissues could begin by 2015. The first human application of bioprinters will probably be to produce tissue for toxicology tests, which will enable researchers to evaluate the effects of drugs on bioprinted models of the liver and other organs.

The coming revolution in 3-D printing brings concerns as well, of course. One of the starkest is the possibility that designs for a printable firearm will be freely available on the Internet. The mission of the group Defense Distributed is to facilitate "a working plastic gun that could be downloaded and reproduced by anybody with a 3-D printer." (The weapon, which the group calls the "Wiki Weapon," would still require conventionally produced bullets.) "We will have the reality of a weapons system that can be printed out from your desk," states Defense Distributed founder Cody Wilson. "Anywhere there is a computer, there is a weapon." While Wilson's manifesto touts quotes by Revolutionary War–era patriots, it's hard to imagine even Patrick Henry not balking at the prospect of a printable gun in the 21st century. No age limits, no background checks, no serial numbers etched on the barrel, no sales receipts, undetectable by security scanners. Print it, use it, melt it down.

Despite the concerns, there will be no stopping what is expected to be the coming juggernaut of 3-D printing. And in perhaps the most futuristic wrinkle so far, the aim of the RepRap Project is to build a printer that can print another printer. In the words of the project's website, "RepRap is humanity's first general-purpose self-replicating manufacturing machine." And who said print is dead?

UNIVERSE OF THINGS *From left: a lampshade by designer Janne Kyttanen; Charles Overy of architectural-modeling company LGM with a miniature of Vail Village; a functional differential gear system from MakerBot; Scott Summit of Bespoke Innovations and a prosthetic limb.*

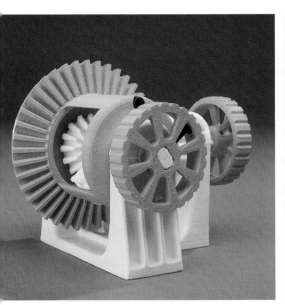

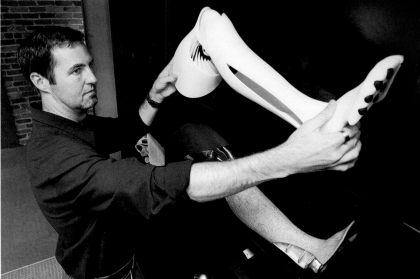

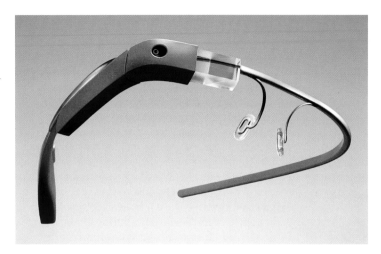

THE VISION THING *Google wants your next pair of glasses to be a computer.*

USERS WILL BE ABLE TO TAKE PHOTOS AND SHARE THEM, **VIDEO-CHAT,** AND ACCESS MAPS AND OTHER INFORMATION ON THE WEB.

Through the Google Glass

Nothing says sci-fi like a set of goggles with special powers. But the aesthetic usually runs to cyborg military or steampunk. The sleekly minimalist Google Glass upends those expectations. Google's next big thing, hatched in the company's famously secretive skunkworks, mounts a tiny transparent rectangle above one eye and holds it in place with trimly asymmetrical frames. What pops up on the screen could be maps, texts, calendars—think personal assistant meets 21st-century monocle. And consider the possible apps. Face-recognition software for the forgetful? Or maybe social-media links to instantly check friends in common?

Technology has a way of eliciting reactions that range from "I've got to have that!" to "Why would I ever need that?" Google's Project Glass, one of TIME's Best Inventions of 2012, is likely to generate both, and perhaps simultaneously. Figuring out what needs might be fulfilled by an eyeglass computer, and what needs might be created, will be a work in progress. And there is also the practical question of whether a tiny floating screen will ever feel, well, natural. The fact that wearing earbuds and Bluetooth headsets is second nature to millions probably argues yes, particularly if computer glasses free us from fumbling with handheld devices. The big question might be: who will be the first to sell ads in that space?

The Honeycomb Skyscraper

China has made it clear it embraces the work of radically innovative ideas, often from architects outside its borders. Who can forget the stunning National Stadium, or "Bird's Nest," unveiled during the 2008 Beijing Olympics, designed by the Swiss firm Herzog & de Meuron and iconoclastic Chinese artist Ai Weiwei? In the years since, China can also boast a surge in homegrown talent. One of the brightest stars among world-class Chinese architects is Ma Yansong, of MAD architects. Ma and colleagues, who recently designed the spectacularly curvy Absolute Towers in Ontario, Canada, are now completing an equally innovative building in Tianjin, China.

The Sinosteel International Plaza is an architectural and engineering marvel because the building's steel exterior ingeniously serves both aesthetic and structural purposes. "The honeycomb façade is what's holding the building up," according to MAD. "The skin is the structure." This eliminates the need for massive internal support structures, freeing up interior space. The honeycomb also enhances energy efficiency. Although the pattern and variable size of exterior hexagons at first appear to be random, they actually follow sun and wind patterns on the building, according to MAD: "By mapping air-flows and solar direction across the site, we were able to position different sized windows accordingly, minimizing heat loss in the winter and heat gain in the summer." Honeycomb architecture challenges conventional construction technology, but as team MAD reminds us, it is a simple concept, found in nature and rooted in a 5,000-year-old architectural tradition.

MORE THAN SKIN DEEP *A honeycomb façade keeps this building standing.*

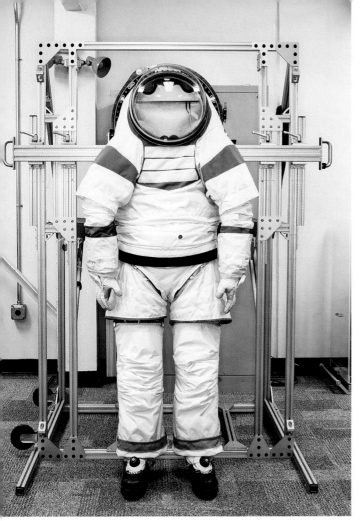

NOT OFF THE RACK *NASA will spend years to perfect its new spacesuit.*

NASA Suits Up for Mars

Spacesuits in the popular imagination are pretty straightforward: climb in, zip up, pop on the helmet, and you're ready to escape the capsule or fight aliens. Reality is more complicated. Designing spacesuits is indeed rocket science. The suits required for walks in space or a stroll on the moon are highly complex systems that require years to develop as well as extensive training to use; the suit is essentially a portable spaceship.

Which is why NASA's new Z-1 spacesuit and portable life-support system (PLSS) is such a great leap for space travelers. The Z-1 is a rear-entry suit with a built-in "suitport" that allows it to dock with a portal on a spacecraft so an astronaut can crawl through and suit up without using a larger airlock that would ordinarily be required—which, in turn, means a big savings of time. The Z-1 also allows for greater freedom of movement.

The PLSS backpack handles all life-support functions, providing oxygen and removing carbon dioxide as well as cooling the astronaut and the suit's electronics. The current prototype uses new technology for its cooling function, designed for the Martian environment. The system that removes CO_2 is also new and will allow longer spacewalks. Both the Z-1 and the PLSS are still in the prototype stage; NASA will spend years refining the systems before they ever see action on another planet.

Liberating Vehicle for Wheelchair Users

Wheelchair-bound car drivers have limited options. Modifying a minivan's controls is expensive, and many people in wheelchairs don't have the upper-body strength to lift themselves in and out of car seats. Stacy Zoern, who has spinal muscular atrophy, is one of them. But Zoern, as CEO of the company Community Cars, is also working to expand those options.

Zoern first learned that a wheelchair-accessible car called the Kenguru was being developed in Hungary, but the company needed financing. Zoern helped raise the money and then quit her job to bring production to Texas. The first Kenguru rolled off the line in January 2012. The car is 100 percent electric, retails for $25,000, has a top speed of 25 mph and a range of 60 miles, and is street legal. And most important, it is built from the ground up to accommodate a wheelchair (the hatchback allows the wheelchair to roll right into the driver's seat).

The current model, however, is made for manual wheelchair users and is driven with a motorcycle handlebar. A model fitted with a joystick and large enough for power wheelchairs is in the works. Zoern, who is waiting for the joystick model herself, is once again raising money as well as hope for would-be drivers.

KENGURU HATCHBACK *Wheelchair users who want to drive a car to get around town now have a new option on the lot.*

FINALLY, A CAR DESIGNED FOR THE

WHEELCHAIR-BOUND.

DRIVERS POP THE TRUNK AND ROLL INTO PLACE BEHIND THE STEERING WHEEL.

Voyage to the Abyss

Filmmaker James Cameron, famous for directing such blockbusters as *Terminator, Titanic,* and *Avatar,* made history as an explorer in March 2012 when he piloted the *Deepsea Challenge* on a solo dive to the bottom of the Mariana Trench, 6.8 miles below the surface of the western Pacific (a depth that would leave Mount Everest still submerged by more than a mile of water). The 24-foot-long submersible, which Cameron helped design, is unique in that it moves through the water in a vertical position, which speeds descent and thereby maximizes the time available to spend on the bottom. U.S. Navy Capt. Don Walsh and Swiss engineer Jacques Piccard made the first descent into the abyss of the trench in 1960 but could stay for only 20 minutes. Cameron hung on for nearly four hours.

The *Deepsea Challenge* was equipped with four HD cameras and an eight-foot array of LED lights. The pilot quarters were so small that Cameron had to keep his knees bent (he practiced yoga to train for the hours of immobility). After his history-making trip, Cameron, who worked with scientists and engineers from several oceanographic-research institutions, appeared to be already planning for a sequel. National Geographic, meanwhile, has movies and television specials in the works that will chronicle Cameron's remarkable underwater adventure.

6.8 MILES

BELOW THE SURFACE OF THE WESTERN PACIFIC LIES THE CHALLENGER DEEP, THE VERY BOTTOM OF THE PLANET'S OCEANS.

CAPTAIN CAMERON *The* Deepsea Challenge *was designed to collect scientific data, specimens, and images from the ocean bottom.*

DREAM WHEELS *A sturdy, light bike so cheap and green it's disposable.*

A 100% Recycled Cardboard Bicycle

Evaluating the quality of a product by saying, "It must be made out of cardboard!" is usually a stinging disparagement. But not to Israeli designer Izhar Gafni, who has devoted years to perfecting what engineers assured him was an impossibility: a cardboard bicycle. Gafni, who had previously designed a pomegranate-juice press and a sneaker-sewing robot, was intrigued with the possibilities of cardboard as a "green" resource. When he found there was very little known about its engineering and structural properties, Gafni had no option but to jump in as a self-made cardboard scientist.

The result, after three years of experimentation and six prototypes, is a lightweight, strong, and durable bicycle made from 100 percent recycled materials. The bike costs about $9 to build and weighs just 20 pounds but can support 480 pounds. It needs no maintenance, has no metal parts, and uses puncture-proof solid tires made of reconstituted rubber from discarded car tires. Gafni estimates the water-resistant bike should last for at least two years of use, even in the rain. There are also plans for child-size bikes and models with detachable electric motors.

Once in production, the innovative bikes could be a fashionable hit in urban centers around the world, but Gafni also hopes that with government rebates for using green materials, the bikes could be given away free in developing countries. Villages of children could all pedal to school on cardboard bikes. And Gafni's vision doesn't end there. He imagines "100 future projects made out of cardboard." A once-humble packaging material is looking good as gold.

THE BIKE WEIGHS JUST 20 POUNDS BUT CAN SUPPORT A WEIGHT OF

480 POUNDS

AND COSTS ABOUT $9 TO MANUFACTURE.

THE NANO AIR VEHICLE, WEARING THE PERFECT DISGUISE, CAN DO JUST ABOUT ANYTHING A REAL HUMMINGBIRD CAN DO.

AERONAUTIC FEAT *This robotic hummingbird can fly in any direction, hover, rotate, and ascend or descend vertically.*

It's a Bird, It's a Plane … It's Surveillance Hummingbird!

Humans are rightly proud of building all manner of planes, jets, and helicopters. But to successfully imitate the flapping flight of birds is actually a much tougher aeronautic challenge. Not that we haven't tried. It's been a dream since the ancient Greek myth of Icarus. Leonardo da Vinci pondered the possibility as well. But only toymakers can claim any degree of success.

Until now, that is. A team of engineers at the California-based company AeroVironment has accomplished the feat of flapping flight and then some, with its Nano Air Vehicle (NAV), a tiny, two-winged surveillance prototype for the Defense Advanced Research Projects Agency (DARPA). Designed to mimic a hummingbird's flight, the NAV can fly up to 11 mph in any direction, even backward. It can hover and rotate clockwise and counterclockwise, it can ascend and descend vertically, and it can fly indoors or out, as well as through open doors. The NAV is controlled remotely like a drone and weighs two thirds of an ounce (that's more than most real hummingbirds, but less than the largest species found in nature). It's also equipped with a video camera. Because it's so small, the NAV can safely go where humans can't. It can scout out combat zones, spy on drug lords hanging out at their tropical cabanas, or hunt for survivors after an earthquake or building collapse.

The technical wizardry of the NAV is stunning, and the biomimicry is Hollywood-clever. On that last point, though, DARPA will need to keep in mind that hummingbirds aren't native to many parts of the world. But perhaps that won't be a problem. The NAV's faux plumage can be changed, perhaps to that of a tiny sparrow (presumably a stunted pigeon would look suspicious). And meanwhile, the flapping robotic menagerie is expanding. Researchers in Japan have devised a beautifully detailed robotic swallowtail butterfly that mimics the soothing and undulating flight trajectory of the real thing.

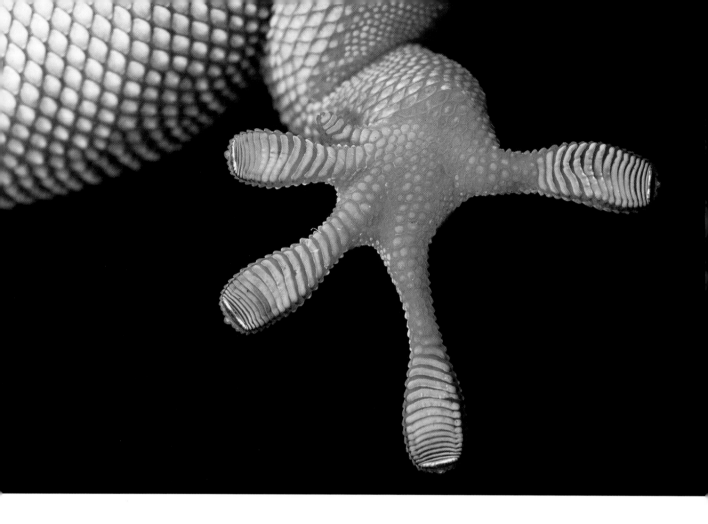

Baby-Soft Bandage Inspired by Gecko Toes

Bandage technology—it sounds like an oxymoron. But imagine pulling bandages off a newborn baby. Removing adhesive strips and tape from such vulnerable skin can not only cause great pain but also be potentially dangerous by stripping away layers of skin. This is why researchers are constantly seeking ways to develop a kinder, gentler, but still effective bandage technology.

Researchers at MIT have long been inspired by the feet of the gecko in their search for better adhesives. The bottom of a gecko's toe is covered with flaplike ridges called lamellae, which are covered by millions of tiny hairs called setae, only a tenth as wide as a human hair. Each seta branches into even smaller strands, like split ends. That microstructure is the basis for the gecko's ability to adhere to smooth surfaces. Unlike glues, however, the gecko's ability to adhere is directional, meaning that a subtle rotation of the toes releases its grip.

One of the promising recent efforts to mimic the quick-release mechanism of the gecko is a bandage with three separate layers: an outside or backing layer, an intermediate layer, and an adhesive that sticks to the skin. The key is that the layers separate rather than stay intact. When the bandage is pulled off, only the adhesive layer stays on the skin, where it gradually wears away. The MIT team hopes the new tape can be tested in clinical settings soon. Meanwhile, the natural adhesiveness of gecko toe pads has also inspired mechanical engineers at Stanford University to design a gecko-like robot (called Stickybot) that can climb up glass. Gecko power indeed.

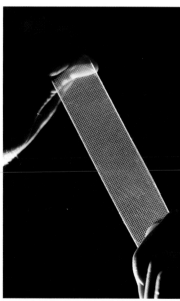

SKIN PROTECTION *A new easy-release bandage could eliminate the pain and tissue damage caused by conventional adhesive tapes. Newborns, the elderly, and patients with injured skin would particularly benefit.*

GECKO STRENGTH: SCIENTISTS HAVE LONG TRIED TO ADAPT THE GRIP PADS ON GECKOS' FEET TO PRACTICAL USE.

Holographic Battle Plan

One of the most important ways for soldiers to prepare for a fight is to visualize the battlefield. With that in mind, the Pentagon's Defense Advanced Research Projects Agency (DARPA) has recently commissioned a major new piece of technology: a holographic table called the Urban Photonic Sandtable Display (UPSD), which shows buildings and terrain in full color and three dimensions. The purpose is to maximize strategic planning and minimize surprises on the urban battlefield.

For millennia, military planners have gathered around bare patches of ground and used sticks to draw plans in dirt or sand. The practice gave rise to the military sand table—basically an elevated sandbox around which officers could strategize. For the most part, improved maps as well as aerial and satellite imagery took over. But even in the modern era, the term "sand table" remains (actual use has waned, though sand-table specialists in some cases still use tons of sand, rock, and gravel to, for example, create large-scale models of terrain in Afghanistan).

The UPSD, however, is definitely a 21st-century model. It was five years in development and created in recognition of the fact that many of today's conflicts take place in urban settings. A team of up to 20 planners can view the holographic display, which, as DARPA hopes, "has opened the door to a new approach to training, mission planning

URBAN STRATEGY *Holograms help planners visualize the battlefield.*

and data visualization." The UPSD is the latest of several holographic tools developed by the military, including maps that have been used in the field in Iraq and Afghanistan. The technology behind the UPSD, which was developed for DARPA by Zebra Imaging, may soon find applications beyond the battlefield.

IT JUST AIN'T SO ...

Tech Gadgets: Not a Boys' Club

If ever there was a sure bet in the technological age, it would seem to be that boys and men are the gadget geeks: the sex that buys, consumes, lives, and dies by the latest in technology. That sure bet turns out to be wrong, however. In a recent survey conducted for HSN by Parks Associates and presented at the 2012 Consumer Electronics Show, it was women, not men,

who were more likely to buy hot tech products.

The study found that not only were women more interested than men in shopping for smartphones, laptops, and tablets, they were also more likely to purchase those items. The only tech category in which men were more interested than women was flat-screen LCD televisions. Women had purchased an average of 4.7 consumer-electronics products in the previous year; men had purchased 4.2 (overall, 88 percent of women had bought a tech-related item, compared with 83 percent of men). Women were also more likely to use social networks and play Facebook games. And when it came to buying a particular tech product, 32 percent of women said ease of use was most important, compared with 26 percent of men. (In a separate study that seemed to confirm rather than upend stereotypes, Gadget Helpline found that 64 percent of male callers had not read the instruction manual before phoning for help, vs. only 24 percent of female callers.) "Women are really the powerhouse in the household driving purchase decisions," Jill Braff, executive vice president of digital commerce for HSN, told the social-media site Mashable.

One study can't be definitive, and there is always the danger of one broad generalization simply being replaced by another. But the newest message is clear: no tech company or retailer that wants to stay in business should underestimate the purchasing power of women.

Chemistry

EDIBLE FOOD PACKAGES ■ SUPER MOLECULAR GLUE ■ IS THERE LEAD IN YOUR LIPSTICK?
SECRETS OF CELL SIGNALS ■ SWEET NEW BATTERIES ■ ADDICTION-PROOF PAINKILLERS
ACIDIC OCEANS ■ VAN GOGH'S COLOR CHANGE ■ THAT OLD-PEOPLE SMELL

The High Cost of Toxic Children

Can a dollars-and-cents accounting of the medical impact of environmental poisons help us see the light?

BY DAVID BJERKLIE

Toxic chemicals in the environment can take a terrible toll on children. And according to public-health experts it's possible—and potentially helpful—to put a price tag on that. The latest estimate, according to researchers, is that the medical impact of environmental toxins on the health of American children is nearly $77 billion a year. It's a staggering statistic, and one that appears to be growing even larger. It was in a landmark study conducted in 2002 that Philip Landrigan and colleagues at Mount Sinai School of Medicine in New York City first estimated the annual cost of four childhood conditions—lead poisoning, cancer, developmental disabilities, and asthma—that could be connected to environmental factors. Landrigan estimated that those factors cost nearly $55 billion, or about 2.8 percent of total U.S. health-care spending. The single largest component of these costs was the diminished lifelong economic productivity that resulted from permanent reduction in cognitive capability in children exposed to neurotoxic chemicals like lead. A decade later, two of Landrigan's colleagues, Leonardo Trasande of Mount Sinai and Yinghua Liu of the National Children's Study, took another look at the costs of environmental disease in children and came up with the $77 billion number.

The conclusions were obvious: in addition to the human costs, environmental pollution and toxins were a significant drag on the economy; the economic case to be made for reducing childhood exposure was clear. There were some glimmers of good news. Trasande and Liu found that exposure to lead and air pollution has been falling, thanks in part to tightening regulations. That has helped reduce mental disability and asthma in children. But it turns out that exposure to toxins may play an under-recognized role

Bisphenol A (BPA)

WHAT IT IS: A chemical used in plastic production

FOUND IN: Water bottles, baby bottles, plastic wraps, food packaging

HEALTH HAZARDS: The government's National Toxicology Program has concluded that there is some concern about brain and behavioral effects on fetuses and young children at current exposure levels.

WHAT YOU SHOULD KNOW: Switch to glass products when possible.

Oxybenzone

WHAT IT IS: A chemical used in cosmetics

FOUND IN: Sunscreens, lip balm, moisturizers

HEALTH HAZARDS: Linked to hormone disruption and low-birthweight babies

WHAT YOU SHOULD KNOW: About 97 percent of Americans have the compound in their urine, but current exposure levels have been deemed safe.

Fluoride

WHAT IT IS: A form of the basic element fluorine

FOUND IN: Toothpaste, tap water

HEALTH HAZARDS: Neurotoxic and potentially tumorigenic if swallowed. The American Dental Association advises that children under 2 not use fluoride toothpaste.

WHAT YOU SHOULD KNOW: Government studies support current fluoride levels in tap water, but studies of long-term exposure and cancers are ongoing.

Parabens

WHAT THEY ARE: Synthetic preservatives

FOUND IN: Products like moisturizers and hair-care and shaving products

HEALTH HAZARDS: Cause hormone disruptions and cancer in animals

WHAT YOU SHOULD KNOW: The FDA has deemed current levels in cosmetics safe, but paraben-free products are available.

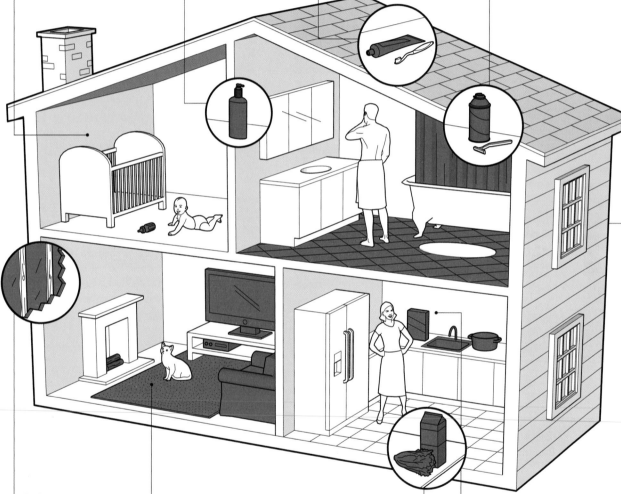

Asbestos

WHAT IT IS: A naturally occurring fibrous mineral

FOUND IN: Housing insulation, drywall, artificial fireplace logs, toys

HEALTH HAZARD: Mesothelioma, a fatal cancer

WHAT YOU SHOULD KNOW: Asbestos in products is not always labeled, and while most manufacturers have abandoned it or reduced its levels, it's still not banned by the U.S. government.

Decabromodiphenyl ether (Deca)

WHAT IT IS: A flame retardant

FOUND IN: Electronics, furniture, carpets

HEALTH HAZARDS: Permanent learning and memory deficits; hearing defects; decreased sperm count in animals

WHAT YOU SHOULD KNOW: Following EPA advice, the industry began phasing out the chemical in December 2009.

Perchlorate

WHAT IT IS: An oxidant in fireworks and rocket fuel

FOUND IN: Drinking water, soil, some vegetables

HEALTH HAZARD: Disrupts thyroid's hormone production

WHAT YOU SHOULD KNOW: Environmental groups are urging the government to lower perchlorate levels in drinking water.

Butylated hydroxyanisole (BHA)

WHAT IT IS: An additive that preserves fats and oils in food and cosmetics

FOUND IN: Chewing gum, snack foods, diaper creams

HEALTH HAZARD: May promote cancer in lab animals

WHAT YOU SHOULD KNOW: BHA is hard to avoid in foods, but the government limits its levels.

in a wider variety of health problems than once imagined.

The impact of toxic chemicals on children begins before birth. Pregnant women who are exposed to high levels of air pollution may be increasing the chances that their children will develop anxiety, depression, and attention problems. In urban areas, nearly everyone is exposed to compounds called polycyclic aromatic hydrocarbons (PAHs), found in car exhaust and cigarette smoke. (PAHs are an example of a class of chemicals known as endocrine disrupters, which mimic naturally occurring hormones and can interfere with developmental and metabolic functions.) A 2012 study found that 100 percent of the New York City women who participated had detectable levels of PAHs in their homes. The researchers, led by Frederica Perera of Columbia University, found that pregnant women with the highest blood levels of PAHs were also more likely to have children with anxiety, depression, and attention problems by age 6 or 7. Women who had the highest levels of PAHs in their homes were four to five times more likely to have children with anxiety problems that might qualify for a clinical diagnosis. "Our study provides new evidence that prenatal exposures to these air pollutants, at levels commonly encountered in New York City and other urban

The Hazards Lurking at Home
A guide to some of the toxins most commonly found in our household products.

areas, may adversely affect child behavior," says Perera.

In an earlier study on factors that affect fetal development, Perera reported that higher levels of PAHs in umbilical-cord blood were linked to more symptoms of anxiety, depression, and attention disorders in children at ages 3 and 5. She is careful to note that "you can't draw conclusions from our results about any single child, or conclude that exposure to PAHs causes behavioral symptoms later. But the results do add to existing evidence that these exposures could have deleterious effects in children."

Behavioral problems are only one aspect of the impact on children of hazardous chemicals in the environment. Other recent studies have found that a woman's exposure to PAHs during pregnancy may also increase the chance of having overweight kids. Columbia University epidemiologist Andrew Rundle and colleagues tracked air-pollution exposure in women in their third trimester of pregnancy and found that children born to mothers with the highest blood levels of PAHs in that trimester had a 79 percent greater risk of becoming obese, compared with children born to moms with the lowest PAH levels. By age 7, the risk was even higher—more than 2.25 times greater.

A good part of what makes environmental toxins so frightening is their insidiousness. A 2012 report concluded that exposure to ubiquitous household chemicals may lower children's responses to vaccines. The study, published in

The Journal of the American Medical Association, suggested that perfluorinated compounds (PFCs), chemicals found in Teflon coatings in pots and pans as well as in furniture, stain-resistant carpeting, rain gear, and microwave popcorn bags, may hinder children's ability to mount proper immune responses after they are vaccinated.

In the report, Philippe Grandjean of the University of Southern Denmark and his colleagues studied children in the Faroe Islands. The researchers chose that population, located in the north Atlantic, since most residents rely on the sea to survive, and recent studies have recorded increasing amounts of PFCs in the drinking water and fish there. All the children received the diphtheria, tetanus, and pertussis (DTaP) vaccine at 3 months, and a booster at age 5. The scientists tested the children's antibodies to diphtheria and tetanus just before they received their booster shot, and again when they were 7. (Antibodies are proteins produced by the immune system in response to foreign invaders and serve as a proxy for immune function.) In addition, the team drew the children's blood to test for PFCs.

When the researchers compared the participants' antibody levels with the levels of PFCs in their blood, they found that higher levels of PFCs were linked with a lower immune response. In fact, kids whose PFC levels were twice as high as other children's had half the amount of antibodies to diphtheria and tetanus as those who had lower blood levels of PFCs. At age 7, the kids with a twofold increase in PFC levels were two to four times more likely to show an immune response that was so low it was no longer clinically protective. "We were kind of shocked when we saw those numbers," says Grandjean. "This is the first study to say that by [exposing children to these chemicals], we are screwing up a major aspect of disease prevention in our society. I've been in the field for quite a while, and this is a very strong signal."

For children who may have been exposed to enough PFCs to affect their immune system, the immediate solution is to get revaccinated. But, says Grandjean, "that can only put a Band-Aid on the wound. The problem is that those drops may not be the only deficit." If the exposure to environmental chemicals is negating the effect of childhood immunizations, then the major public-health gains made by vaccines in the past century in preventing infectious diseases may start to erode. And if the immune system is compromised enough to respond more sluggishly to vaccines, what does that mean for its ability to fend off other pathogens, like the common cold virus and influenza? Further, is it possible that exposure to environmental chemicals plays a role in autoimmune diseases, or conditions like type 1 diabetes and cancer, that may in part be due to dysfunction of the immune system?

There's not much we can do currently to cleanse ourselves of these compounds, says Grandjean. But if we're aware of how potentially harmful they can be to our health, we might be more vocal about how we want our regulatory agencies to handle them. "When we see results like this, it's clear we haven't done our job well enough," he says. "I think the next generation deserves better from us."

How About a Second Helping of That Container?

INSTEAD OF MAKING
MOUNTAINS OF TRASH
WITH THE PACKAGES THAT FOOD COMES IN, NOW WE CAN EAT WHAT WE ONCE THREW AWAY.

Imagine a dinner party (OK, it's actually a nightmare) in which guests eat every bit of food in your refrigerator, freezer, and pantry. It gets worse. Then imagine the enormous pile of plastic, paper, glass, foam, cardboard, and metal containers left behind. In reality, food packaging is scary because we dump millions of tons of the stuff into landfills every year.

Harvard bioengineering professor David Edwards and French designer François Azambourg aim to change that, however, with packaging you eat. The edible containers, called WikiCells, can be used to wrap foods like ice cream or yogurt or liquids like orange juice or hot chocolate. The wrapper is a tasty skinlike membrane made of food particles such as dried fruit, nuts, and seeds that are held together by calcium or magnesium ions and alginate, derived from seaweed.

An outer package will also be required in most cases and will be made from ingredients such as tapioca, caramel, and bagasse, a byproduct of sugar-cane processing. Some versions of the outer package will be edible, though it's debatable how many people will want to chew on food boxes. In all cases, the outer packaging will be easy to compost.

In perhaps the best sign of the strength of the edible-packaging trend, other companies are getting into the act too. Loliware makes flavored, edible glasses; MonoSol makes water-soluble films to wrap preportioned pouches of drink powders.

EAT, DRINK ... AND EAT
Munchable glasses from Loliware (above) are an alternative to disposable plastic cups; natural wrappers from WikiCells (left) can enclose ice cream , yogurt, or cheese and eliminate conventional packaging.

Superglue From Bacteria

Scientists have devised a natural superglue from a most unusual source: *Streptococcus pyogenes,* a bacterium that causes 700 million infections each year. The pathogen usually causes only mild to moderate illness, including strep throat, but in rare cases can lead to necrotizing fasciitis, a terrifying and life-threatening infection that has earned *S. pyogenes* infamy as "the flesh-eating bacteria."

Oxford University biochemists Mark Howarth and Bijan Zakeri created their superglue by harnessing the protein central to *S. pyogenes*'s infectious arsenal, a protein characterized by an intramolecular bond that is virtually inseparable. The researchers devised a way to split the protein and separate the two parts, dubbing one of the fragments SpyCatcher and the other SpyTag— "Spy" being lab slang for *S. pyogenes.*

This won't be the kind of glue that repairs your broken coffee mug (at least not yet), and it won't stick to your fingers (the SpyCatcher and SpyTag fragments will stick only to each other). The first use will probably be in laboratories, where scientists find molecular glues useful in research. Future applications, however, may include building custom enzymes on an industrial scale.

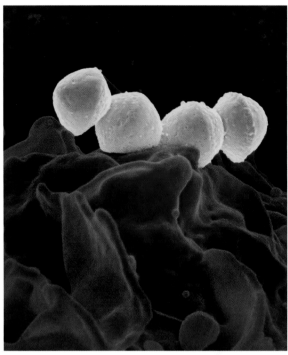

STUCK ON YOU S. pyogenes *attacks with a natural superglue.*

400
POPULAR LIPSTICKS CONTAIN TRACE AMOUNTS OF LEAD, ACCORDING TO THE FDA.

Is There Lead in Your Lipstick?

Your favorite shade of fire-engine red may contain lead, according to the U.S. Food and Drug Administration (FDA). The agency found that 400 popular lipsticks contained trace amounts of the toxin. The worst offenders on the list were Maybelline's Color Sensational in Pink Petal, which had 7.19 parts per million of lead, and L'Oréal Colour Riche in Volcanic, which had seven parts per million. Several other brands, including Cover Girl and Nars, had products hovering in the four-to-five-parts-per-million range; those figures are more than double the highest levels the FDA found in its first lipstick-lead test in 2007, which looked at 20 lipsticks and found lead in all.

The FDA began testing for lead in lipstick in response to pressure from the consumer organization Campaign for Safe Cosmetics. The concern, said the group, is that "lead builds up in the body over time and lead-containing lipstick applied several times a day, every day, can add up to significant exposure levels"—especially worrisome for women of childbearing age, since exposure to lead during pregnancy can interfere with fetal development.

Lead is never intentionally added to cosmetics, says an industry trade group. And although the FDA does not believe that lipstick lead poses a safety concern, it has acknowledged it is "evaluating whether there may be a need to recommend an upper limit for lead in lipstick in order to further protect the health and welfare of consumers." With levels of lipstick lead seemingly on the rise, a solution that is more than cosmetic might be in order.

Hormones or neurotransmitters first bind to a cell's surface receptor; proteins (purple) then turn on the cell's internal machinery.

10

TRILLION CELLS
IN THE HUMAN BODY
STAY CONNECTED
THROUGH A SYSTEM
OF RECEPTORS FOUND
IN CELL WALLS.

Unraveling the Mysteries of the Body's Internal Signals

Our bodies face a staggering communication challenge. Trillions of cells, with hundreds of specialized forms and functions, must send and receive critical information every second we are alive. This constant exchange of information is mediated through microscopic sensors called receptors, located on the surfaces of cells. Teasing out how these receptors work earned researchers Robert Lefkowitz of Duke University and Brian Kobilka of Stanford University the 2012 Nobel Prize in chemistry.

Scientists have long known that signaling molecules like hormones course through our blood and trigger changes in our bodies. When we are stricken with fear, for example, nerve signals from the brain fire a warning, which causes the pituitary gland to release hormones that awaken the adrenal gland, which starts pumping out cortisol, adrenaline, and noradrenaline; blood is flooded with sugar and fat, the lung's bronchi expand, and the heart races in order to boost oxygen and energy levels to muscle cells. But how do hormones streaming through the blood activate changes within cells? We now know that receptors embedded within cell walls relay the signals to internal cellular machinery. But finding them and understanding their workings was a long process.

The first step was to pinpoint receptors within cell walls. Kobilka and Lefkowitz then identified the genetic blueprint for a receptor, and it became clear there were other receptors that looked alike and functioned in the same manner. The last piece of the puzzle was to take a molecular "snapshot" of the receptor in action, a stop-action record of the hormonal key turning the cellular lock.

Researchers have since determined there are about 1,000 genes that code for such receptors, which transmit the cellular instructions carried by a number of different hormones as well as convey the signals that create our sensory experiences of light, flavor, and odor. Unraveling the workings of this class of receptors is fundamental to understanding our physiological communication system. The discoveries also have a wide range of practical implications, since these same receptors and signaling mechanisms are the ones that mediate the action of roughly half of all pharmaceutical drugs.

Sweet! Sugar-Powered Batteries

Researchers are working hard to devise an alternative to the rechargeable lithium-ion battery, which is the current standard in consumer electronics and is increasingly considered the battery of choice for electric cars. The problem is not so much performance as price and availability; lithium is expensive and relatively scarce. Though there's little chance the world will run out of lithium, most of it is mined in high-altitude desert salt flats in northern Chile and Argentina. A worldwide fleet of electric cars could put pressure on that supply.

Researchers at Tokyo University of Science in Japan, a country that's a huge consumer of lithium batteries, report they are making significant progress on an alternate battery design that uses sodium ions (available from many sources, including table salt) and a hard black carbon extracted by superheating sucrose (derived from sugar) at temperatures of up to 2,700°F in an oxygen-free atmosphere.

The team anticipates it could take about five years to achieve a practical version of its sodium-sucrose design, but the group is confident this is possible. If so, it could mean batteries made from materials that are essentially unlimited in supply.

That doesn't mean, however, that an ultimate battery will be here anytime soon. Batteries have come a long way since Ben Franklin experimented with a rudimentary version some 250 years ago, but progress has been incremental rather than exponential. In many otherwise high-tech products (think laptops and cell phones), batteries remain the weak link. Perhaps those made from sugar and salt can help pave the way for future improvements.

THE NUMBER OF BATTERIES WE BUY EACH YEAR,
3 BILLION,
IS A SIGN WE NEED A BREAKTHROUGH.

An Addiction-Proof Painkiller?

The problem with painkillers is that they often work too well—they relieve pain but also produce a high that proves irresistible to many people. But researchers believe they may have found a way out of this double bind through a fundamental property of certain molecules: chirality. A example of chirality is right and left hands; we think of them as identical, but you can't wear a left-hand glove on your right hand. The same thing occurs in some molecules, and it turns out that the "handedness" of a molecule can affect its physiological properties. When rodents were given morphine plus a drug called (+)-naloxone, a mirror-image molecule of the drug naloxone (which is currently used as an antidote in cases of narcotic overdoses), the animals did not develop the typical signs of addiction. And even better, (+)-naloxone appeared to enhance morphine's pain-relieving effects. The research is still preliminary, and the quest for non-addictive opioids is one of repeated failure—heroin and Oxycontin were once heralded as solutions. But researchers are hoping they have found new possibilities for addiction-free pain relief in the molecular mirror.

Acidifying Oceans Break a Long, Long Record

Human beings don't think in geologic time. And yet to understand our impact on the planet, we need to consider changes that used to take place over millions of years.

For example, we know the oceans are becoming more acidic because of rapidly rising levels of atmospheric carbon dioxide (much of the excess CO_2 pumped into the atmosphere ends up in the oceans, where it dissolves and combines with water molecules to form carbonic acid). But in a review of hundreds of paleoceanographic studies, scientists now believe that the rate of change in ocean acidity is occurring faster than it has for 300 million years. And the pace of acidification will likely only accelerate in the next century. That could have dire effects on marine life. Increasingly acidic seawater directly affects tiny sea creatures such as coral and plankton that have

carbonate shells, which can simply dissolve in more acidic water. When organisms like plankton disappear, species higher up in the food chain are affected too.

This is what's so scary about rapid climate change: we are altering the planet at a breakneck pace, by geological standards at least. Life can adapt, but the faster the planet changes, the harder it will be for countless species—including us, potentially—to keep up. And we have a word for what happens when species can't keep pace with environmental change: extinction.

OCEANS ARE GROWING ACIDIC MORE RAPIDLY THAN THEY HAVE IN

300 MILLION YEARS.

OBSCURED BEAUTY *The chemistry of paint and varnish altered a masterpiece.*

Why a Van Gogh Changed Colors

The ancients claimed, "Ars longa, vita brevis" (Art is long, life is short), but museum conservators know better, especially when it comes to paintings and pigments. For example, some bright yellows in Vincent van Gogh's *Flowers in a Blue Vase* have mysteriously changed to orange-gray over time.

Ironically, the culprit turns out to have been a chemical reaction between the paint and a protective varnish applied to the painting after the artist's death. Varnish becomes brown with age and thereby darkens colors, but it can often be carefully removed, and underlying colors remain vibrant. Not so with *Flowers*—and now scientists know why, thanks to a microscopic analysis conducted at the European Synchrotron Radiation Facility in France and the Deutsches Elektronen-Synchrotron in Germany.

Instead of finding the crystalline cadmium-sulfate compounds that researchers would expect due to normal oxidation of the cadmium-yellow paint Van Gogh used, what they found was a lead-sulfate compound. The apparent source of the lead was a drying agent added to the varnish. The reaction between pigment and varnish produced a crust that cannot be removed without affecting the underlying chromium-yellow paint.

Hmmm... Is That Eau de Elderly?

It seems pretty indelicate, something only a 3-year-old would say out loud, but old people, well, smell old. Don't they? Yes, according to researchers at the Monell Chemical Senses Center in Philadelphia, "old-person smell" is real and identifiable, and it's not just due to mothballs or a musty house. But surprisingly enough, in blind testing, people don't find it an unpleasant smell.

Researchers collected body-odor samples from individuals representing three different age groups; samples were taken from pads sewn into the armpits of T-shirts and then placed into sealed jars. Volunteers were asked to rate the intensity and pleasantness of each odor. Old-person smell was indeed deemed distinctive, but was rated as less intense and less unpleasant than body odor from younger donors (honors for worst-smelling went to middle-aged men).

The root cause of old-person smell is still a mystery, but what seems certain is that our human ability to discern age through scent has evolutionary origins. We may not be as dependent as other animals on our noses, but our brains are still wired to extract useful information from the way other folks smell.

OUR ABILITY TO DETERMINE A PERSON'S
AGE THROUGH SCENT
IS ONE WAY OUR BRAINS COLLECT
INFORMATION ABOUT OTHER HUMANS.

IT JUST AIN'T SO ...

When Germ-Free Bites Back

It sometimes seems that one of the core missions of civilization is to wipe out germs—any and all germs. Exhibit A: the ubiquity of hand sanitizers in our homes, offices, cars, and purses. But while we understandably want to maintain the upper hand in the war against infectious disease, researchers know that cleaner isn't always better. A recent study has found that triclosan, a common chemical ingredient found in antibacterial products, impairs muscle function in lab and animal tests. And this is only the latest study that has implicated the chemical, which has also been linked to promoting bacterial resistance to antibiotics, increasing the incidence of allergies, and disrupting hormone function.

The evidence is still preliminary, however, and as the FDA states, "triclosan is not currently known to be hazardous to humans." And yet the agency adds that in light of recent studies, it is "engaged in an ongoing scientific and regulatory review of this ingredient."

Sounding a common-sense note, the FDA also observes it has no evidence that triclosan provides extra health benefits over ordinary soap and water. In a society bent on creating surgically sterile conditions everywhere, consumers should feel empowered to stop and just say no to the aggressive marketing of antibacterial products.

Genetics

AUTISM'S RANDOM SIDE ▪ EXERCISE YOUR DNA ▪ DROUGHT-RESISTANT CROPS
THE SPIT TEST ▪ BREAKING DOWN BREAST CANCER ▪ THE FETAL BLUEPRINT ▪ HEALTHY GERMS
WHEN SPERM GROWS OLD ▪ THE MYSTERY OF CHILDHOOD TUMORS

Junk DNA—
Not So Useless After All

*Turns out many diseases are triggered not just in single genes
but in a variety of hormones, enzymes, and other metabolic factors.*

BY ALICE PARK

Junk. Barren. Nonfunctioning. Dark matter. That's how scientists had described the 98 percent of the human genome that lies between our 23,000 genes, ever since our DNA was first sequenced about a decade ago. The disappointment in those descriptors was intentional and palpable.

For so long, we believed that the human genome—the underpinnings of the blue-print for the talking, empire-building, socially evolved species that we are—would be stuffed with sophisticated genes, coding for critical proteins of unparalleled complexity. But when all was said and done, and the Human Genome Project finally mapped the entire sequence of our DNA in 2001, researchers determined that humans have only about 23,000 active genes, which made up a paltry 2 percent of the entire genome. The rest, geneticists acknowledged with unconcealed embarrassment, was an apparent bio-logical wasteland.

But it turns out they were wrong. In an impressive series of more than 30 papers pub-lished in journals including *Nature, Genome Research, Genome Biology, Science,* and *Cell,* scientists now report that these vast stretches of seeming "junk" DNA are actually the seat of crucial gene-controlling activity—and changes within them contribute to hun-dreds of common diseases. The new data come from the Encyclopedia of DNA Elements project, or ENCODE, a $123 million endeavor begun by the National Human Genome Research Institute in 2003, which includes 442 scientists in 32 labs around the world.

ENCODE has revealed that some 80 percent of the human genome is biochemically active. "What is remarkable is how much of [the genome] is doing at least something. It

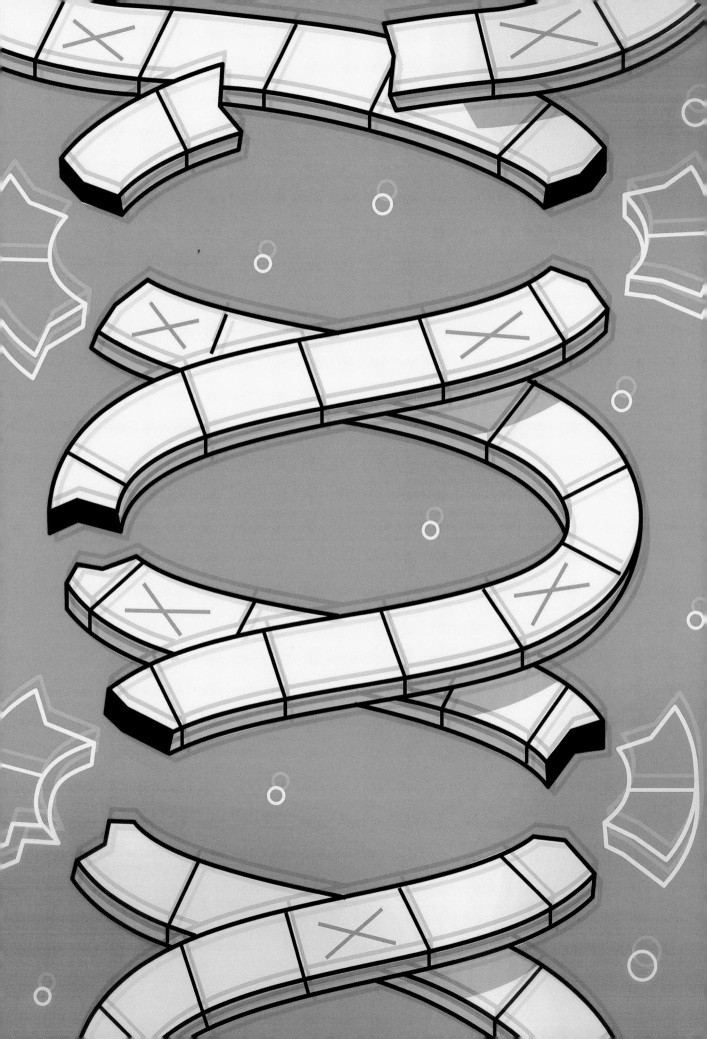

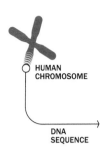

HUMAN
CHROMOSOME

DNA
SEQUENCE

THEN

The genes Making up only about 2% of the genome, these provide instructions for building the body's tissues

The junk Some of the 98% help regulate genes, but the rest didn't seem to make proteins or have any other function

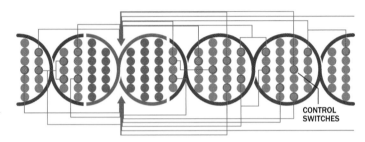

CONTROL
SWITCHES

Junk No More.
The vast majority of the human genome that scientists had written off is actually alive with activity

NOW

The "junk" turns out to be a series of switches that work together to issue instructions to the genes. Manipulating the switches could lead to new cures or treatments

has changed my perception of the genome," says Ewan Birney, ENCODE's lead analysis coordinator from the European Bioinformatics Institute.

Rather than being inert, the portions of DNA that do not code for genes contain about 4 million so-called gene switches, transcription factors that control when our genes turn on and off and how much protein they make, not only affecting all the cells and organs in our body but doing so at different points in our lifetime. Somewhere amid that 80 percent of DNA, for example, lie the instructions that coax an uncommitted cell in a growing embryo to form a brain neuron, or direct a cell in the pancreas to churn out insulin after a meal, or guide a skin cell to bud off and replace a predecessor that has sloughed off.

"What we learned from ENCODE is how complicated the human genome is, and the incredible choreography that is going on with the immense number of switches that are choreographing how genes are used," Eric Green, director of the National Human Genome Research Institute, said during a teleconference discussing the findings. "We are starting to answer fundamental questions like what are the working parts of the human genome, the parts list of the human genome and what those parts do."

If the Human Genome Project established the letters of the human genome, ENCODE is providing the narrative of the genetic novel by fashioning strings of DNA into meaningful molecular words that together tell the story not just of how we become who we are but of how we get sick as well.

Ever since the human genome was mapped, scientists have been mining it for clues to the genetic triggers, and ultimately the treatments, for a variety of diseases—heart disease, diabetes, schizophrenia, autism, to name just a few. But hundreds of so-called genome-wide association studies (GWAS) that have compared the DNA of healthy individuals with those with specific diseases revealed that the relevant changes in DNA were occurring not in the genes themselves but in the noncoding genetic black holes. Until now, researchers didn't fully understand what these noncoding regions did; if variations in these areas were not part of a known gene, they couldn't tell what impact, if any, the genetic change had. And more important, they couldn't begin to design ways of correcting or compensating for the changes.

ENCODE, however, now provides a map of those genetic switches that will allow scientists to determine exactly what those variants do; it's likely that their function in regulating and controlling key genes can now be traced and studied—and hopefully manipulated to treat whatever diseases they contribute to. "We need to revisit the interpretation of those studies," Dr. John Stamatoyannopoulos, associate professor of medicine and genome sciences at the University of Washington, said, referring to the hundreds of GWAS now published. "In many cases those studies concluded that 10 or 15 variants might be important for a particular disease. ENCODE data points to the fact that this is probably a significant underestimate, that there may be dozens, even hundreds of variants landing in switches, so there is a tremendous amount of information still hidden within those studies that needs to be reanalyzed in the context of the new data."

Eager to put their newfound scientific knowledge to work, scientists have already begun some of those studies. At UW, Stamatoyannopoulos and his colleagues found that gene changes identified by GWAS as being involved in 17 different types of cancer seem to affect nearly two dozen transcription factors that translate raw DNA into the RNA that turns into functional proteins. This common molecu-

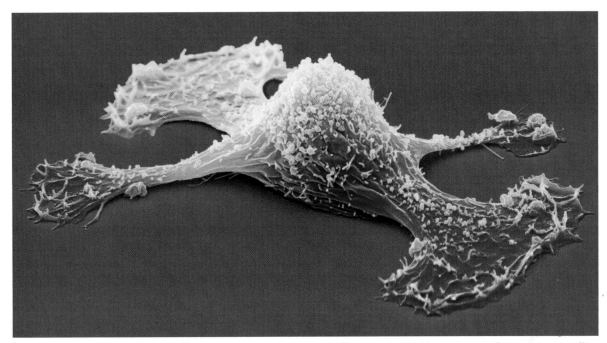

COMMON THREAD *Understanding gene changes seen in 17 types of cancer may lead to new treatments (above, metastasis of an ovarian-cancer cell).*

"We are going to work out how we make humans," says an ENCODE scientist, "starting from the simple instruction manual."

lar thread may lead to new treatments that control the function of these transcription factors in not just one but all 17 cancers, including ovarian, colon, and breast diseases. "This indicates that many cancers may have a shared underlying genetic predisposition," he said. "So we can make connections between diseases and genome-control circuitry to understand relationships where previously there was no evidence of any connection between the diseases."

ENCODE may shed significant light on our most common chronic diseases, including diabetes, heart disease, and hypertension, which result from a complex recipe of dysfunction, not just in single genes but in a variety of hormones, enzymes, and other metabolic factors. Changes in the way some genes are turned on or off may explain the bulk of these conditions, and ultimately make them more treatable. "By and large, we believe rare diseases may be caused by mutations in the protein- [or gene-]coding region," said Green, while the "more common, complicated diseases may be traced to genetic changes in the switches."

In another example of ENCODE's power, Birney said the genetic encyclopedia has also identified a new family of regulators that affect Crohn's disease, an autoimmune disorder that causes the body's immune cells to attack intestinal cells. The finding could lead to novel, potentially more effective therapies. "I've had more clinical researchers come to my door in the past two years than in the previous 10," Birney said. "It's going to be really good fun producing lots of insights into disease over the next couple of years."

Not only does ENCODE open doors to new therapies, it pushes our basic understanding of human development to new heights as well. At the heart of many genetic researchers' investigations is the desire to understand how each cell in our body, from those that make up our hair to those that reside in our toenails, can contain our entire genome yet still manage to look and function in such widely divergent ways. ENCODE's scientists knew that certain regulatory mechanisms dictated when and where certain genes were expressed, and in what amount, in order to give rise to the diversity of cells and tissues that make up the human body, but even they were surprised by just how intricate the choreography turned out to be. "Most people are surprised that there is more DNA encoding regulatory control elements, or switch elements for genes, than for the genes themselves," says Michael Snyder, director of the center for genomics and personalized medicine at Stanford University and a member of the ENCODE team.

In keeping with the open-access model established by the Human Genome Project, ENCODE's data is available to researchers for free on the consortium's website. The database will fuel a renewed interest in genome-based approaches to diagnosing and treating disease. Despite initial excitement in the field, in the years since the genome was mapped, gene-guided treatments and gene-therapy approaches to treating disease have proved difficult to bring to the clinic; part of the challenge, geneticists say, may have been related to the fact that they didn't fully understand how to control the genes that were affected by disease. "I am pretty sure this is the science for this century," Birney says. "We are going to work out how we make humans, starting from the simple instruction manual." And perhaps we'll figure out how to make humans healthier as well.

Understanding Autism: Don't Blame Mom and Dad

It's not genetic changes handed down from parents, but rather random changes in genes that may be responsible for some cases of autism, according to a series of studies that represent a breakthrough in understanding the causes of the developmental disorder.

The findings suggest that autism is a genetically complex condition, involving perhaps hundreds of spontaneous changes in genes. Where past research on the genetic roots of autism looked at families with at least two affected siblings, in which inherited genetic mutations may play a more prominent role in the disorder, the latest analyses focused on families with only one child affected by autism. By comparing the genetic differences between these children and their nonautistic family members, they found that children with autism were about four times more likely than their unaffected siblings to have copy-number variants (CNVs), or mutations in which a part of the genome is either duplicated or deleted. While rare, these CNVs occur in a region of the genome known to be involved in social behaviors and empathy, and thus may provide new targets for potential treatments for autism. "The advances and new information are coming fast and furious now," says Dr. Matthew State of Yale University School of Medicine, one of the study's authors. "It lays an important foundation for the next step, which is to identify better treatments for folks with autism."

How You Can Pump Up Your DNA

Yes, being physically active comes with a host of good-for-you benefits, including keeping you fit and strengthening your heart and muscles. But there's a surprising new plus to add to the list: physical activity can change your DNA for the better.

The basic genetic stuff that makes you you—the unique molecular code that dictates whether you have brown hair or blue eyes or a propensity for cancer—doesn't change. But how you live your life—the things you eat, what you're exposed to in the environment—can affect what's known as your epigenome, the system that determines which genes are turned on at specific times and how long they're active. The key regulators in that process are molecular gene-jumpers known as methyl groups, a type of hydrocarbon derived from methane, which sit on top of genes and decide whether the genes are active or inactive.

In order to understand how methyl groups are affected by exercise, scientists at the Karolinska Institute in Stockholm asked volunteers to exercise at varying levels of exertion on a stationary bike and took biopsies of muscle both before and after the sessions. They found that even a single 20-minute workout can lead to epigenetic changes that help muscles work better, priming them to churn out enzymes and proteins that energize the muscle.

The study represents an advance for the field of epigenetics and reveals the intricate choreography that methyl groups follow in dancing around a muscle cell's genome. And the results only serve as additional proof that the human body was built for working out—right down to its genetic roots.

RESEARCHERS SAY THAT EVEN A SINGLE

20 MINUTE

WORKOUT CAN LEAD TO EPIGENETIC CHANGES THAT HELP MUSCLES WORK BETTER.

Are Superseeds an Answer to Drought?

Nothing distresses a farmer more than helplessly seeing his acres of crops withering away as the soil is wrung dry of all moisture. While farmers know that they'll have to endure the occasional dry spell, recent droughts in the American Midwest have forced growers to confront a new reality: the weather isn't in their control, but the types of crops they grow are. What if there were a way to breed thirsty plants like corn so that they can use water more efficiently during times of drought?

That's what agribusiness is hoping to achieve with new genetically modified (GM) crop strains that contain drought-resistant genes and are designed to welcome arid conditions. Industry leader Monsanto is working on a hybrid line of corn called DroughtGard, developed with the German firm BASF, that is intended to enhance crop yield in dry soils. DroughtGard is the first GM crop approved by the U.S. Department of Agriculture to focus on drought tolerance and features a bacterial gene that enables it to better retain water. Hundreds of farmers in the western end of the Corn Belt—an area that runs to dry even in normal years—are field-testing DroughtGard; Monsanto says early results indicate that the GM crop might improve yields by 4 to 8 percent over conventional crops in some arid conditions. "This year magnifies how important it is to have drought tolerance," says Robert Fraley, Monsanto's chief technology officer.

Still, critics are skeptical that GM crops alone will enable farmers to overcome persistent drought. In a June report, the Union of Concerned Scientists (UCS) noted that GM crops take years to develop and that the seeds created so far offer only modest benefit. "Genetic engineering is not a silver bullet," says Doug Gurian-Sherman, a senior scientist at UCS and the author of the report. If droughts become more frequent, though, farmers may need all the bullets they can get.

GENETICALLY MODIFIED CROPS MAY IMPROVE YIELDS UP TO

8%

OVER CONVENTIONAL CROPS IN SOME ARID CONDITIONS.

What Your Spit Can Tell About You

If you're a fan of *CSI*, you know that dead men tell no lies. Not even about their age.

But it wasn't until a new discovery by researchers at the University of California, Los Angeles, that scientists could actually determine a person's age using genetic material. Despite the sophistication of the latest tools of forensic science, even the cleverest coroner couldn't definitively establish the age of a specific tissue sample or cell.

UCLA researchers report that working with saliva samples, they could find so-called epigenetic alterations to DNA that are caused by environmental influences like diet, stress, exposure to sunlight, and carcinogens and even toxins. The resulting alterations don't change the DNA itself but are layered on top of the genome, affecting how genes are turned on or off. At specific areas of the genome, these changes increase or decrease in almost chronological fashion, allowing them to serve as a timeline to predict a person's age to within five years.

Forensics would be only the first area of applications of this epigenetic bonanza. The scientists also envision being able to study aging outliers who seem to have "younger" epigenomes compared with their chronological age, to determine if they possess some biological advantage to longevity. That might lead to treatments or interventions for the outliers on the other end—those whose cells seem older than their chronological age.

LAB WORK *The applications could go well beyond forensics.*

BY ANALYZING TRACES OF SALIVA LEFT IN A BITE MARK OR ON A COFFEE CUP, EXPERTS COULD NARROW THE AGE OF A CRIMINAL SUSPECT TO A

FIVE-YEAR RANGE.

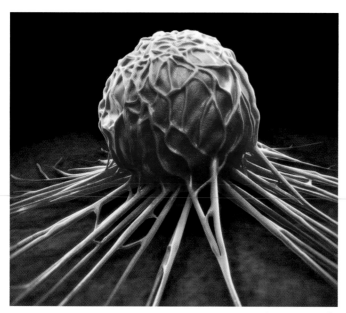

FIRST STEP TO NEW TREATMENTS *A breast-cancer cell.*

Breast Cancer's Four Types

Breast cancer is a complex disease, driven by many genetic and lifestyle factors. But in the latest analysis of the DNA of breast tumors, researchers were heartened to find that the disease may actually be slightly simpler than scientists had thought. As part of The Cancer Genome Atlas (TCGA)—a government project that is aiming to sequence tumor genomes from dozens of different cancers to help scientists better understand tumor development and treatment—scientists sequenced 510 tumors from 507 patients with breast cancer. All told, they found 30,626 mutations in these cancer cells, but those aberrations fell into four main groups.

The groupings open new avenues toward treating the disease, since they forge links between certain breast tumors and other cancers. One subtype, for example, shares genetic similarities with ovarian cancer, leading to the possibility that treatments already in existence to combat that disease might also help women with breast cancer. Another subtype distinguishes women who respond better to certain anti-cancer therapies, which could potentially improve their survival rates, since doctors would know which treatment works best for them. "We are really getting at the genetic roots of these different types of breast cancer," says Charles Perou, professor of genetics and pathology at the University of North Carolina at Chapel Hill and one of the many investigators involved in the project. "The work shows that, really, these four groups of [breast cancer] require major attention as an important first step down the road to individualized medicine. We now have the blueprint for breast-cancer genetics; it's the most complete blueprint ever achieved." And one that could lead to the most effective way of treating cancer as well.

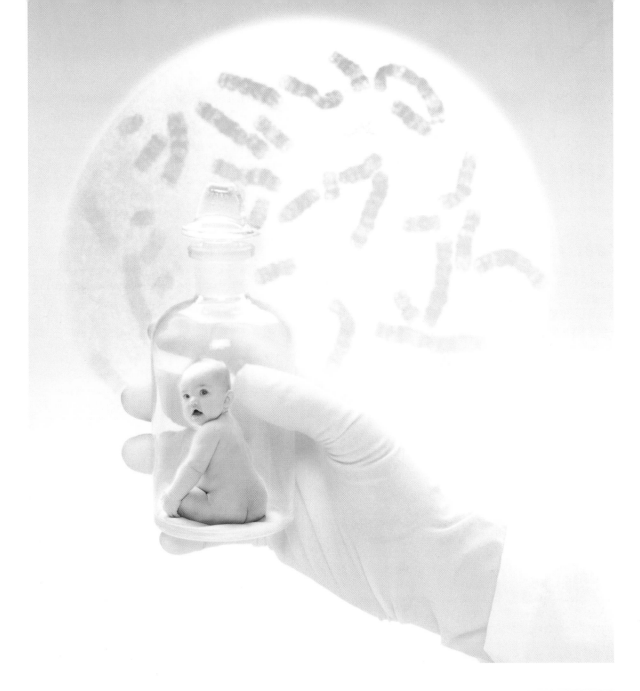

Reading an Unborn Baby's DNA

Suspended in the blood of a pregnant woman, along with some added information from a dad-to-be's saliva, lurks enough fetal DNA to map out an unborn baby's entire genetic blueprint. In June 2012, researchers reported on the first genome sequencing of a baby who was still in utero—a step that takes prenatal testing to new heights, promising a motherlode of genetic information collected without the use of an invasive procedure like amniocentesis. And it comes with a corresponding trove of data that even experts don't yet know how to interpret.

The power of genetic information lies in its predictive capabilities—in its ability to lay out, in DNA, the current and future disease tendencies in a human being. But interpreting such a genetic dossier is still more art than science. While some conditions, like cystic fibrosis, can be traced to changes in a single gene, most diseases are likely the result of multiple DNA changes that don't announce themselves as an obvious network. And the mere presence of a genetic change associated with a disease doesn't necessarily mean an unborn child will develop that condition, which leaves the utility of such testing in doubt, at least for now. Jacob Kitzman, lead author of the pioneering study from the University of Washington, acknowledges the hurdles that genome sequencing faces before it becomes a mainstay in clinics: "It's a really big challenge for the field, figuring out how to communicate to clinicians not only the results but the uncertainty that goes along with those results."

RESEARCHERS FOUND PRENATAL GENOME SEQUENCING WAS MORE THAN

98%

ACCURATE WHEN COMPARED WITH ACTUAL POSTNATAL RECONSTRUCTION.

A Census of the Germs Within Us

The human genome is made up of about 23,000 genes. Which is a fairly impressive figure until you consider this: the number of nonhuman genes each of us carries around—from the bacteria, viruses, and other pathogens living in and on us—tops 8 million.

Indeed, bacterial cells outnumber human cells 10 to 1. That's why the exploration of the human microbiome— the collective population of all the nonhuman cells and genes that inhabit us—is one of the fastest-rising fields of medical research. The National Institutes of Health in 2007 launched the Human Microbiome Project, a multi-year, $153 million effort to sequence the genes of all the microbes we carry with us.

In research published in June 2012, the first results of that effort offered an unprecedented peek into this hidden world of the human microbiome. The scientists described the full genetic sequences of certain familiar species of microbes that reside in various parts of the body, including the nose, mouth, intestines, and vagina, as well as descriptions of bacteria that have never before been seen.

And these microbes are far from being just freeloaders or invaders. Rather, they're crucial facilitators of many of our basic bodily functions: from digesting food and producing vitamins to fending off harmful infections and recovering from illness. They not only keep people healthy, but they may also explain differences in individual health—why people respond differently to the same drug, or why some develop chronic diseases and others don't. Intriguing research also suggests that the makeup of these microbes in our gut, for instance, may influence things like weight or our propensity to develop allergies.

All of which argues for the fact that we should stop stereotyping all bacteria as bad. There are good guys among them, and we should learn to become their allies.

Older Dads and the Risk of Autism and Schizophrenia

Older men are more likely than younger ones to have children with autism or schizophrenia, and a new genetic study points to why: compared with younger dads, older fathers pass on significantly more random genetic mutations to their offspring that increase the risk for these conditions.

And unlike the genetic contributions of the mother, older fathers are responsible for nearly all of a child's random genetic mutations: a father's age at conception may account for 97 percent of the new, or de novo, mutations found in his offspring, according to the new study from deCODE Genetics in Iceland.

The findings may partly explain the rise in autism diagnoses in recent decades, and they shore up previous studies indicating children born to older dads are more likely to have developmental and psychiatric disorders. Not all such mutations are harmful, however, and even those that are must occur in the right combinations to generate disease. "The observed effect is a significant one but not one necessarily to cause great worry among prospective older fathers," Darren Griffin, a professor of genetics at the University of Kent, wrote in a comment accompanying the study's publication. "The numbers of mutations detected in this study are in the dozens...and not realistically likely to deter more mature fathers from having children." It might, however, give them pause about putting off fatherhood for too long.

A FATHER'S AGE AT CONCEPTION MAY ACCOUNT FOR

97%

OF THE NEW, OR DE NOVO, MUTATIONS FOUND IN HIS OFFSPRING.

Decoding Childhood Tumors

Childhood-cancer experts are hoping that at least some of what drives pediatric cancers is hidden in the genomes of patients and their tumors. That's the thinking behind the Pediatric Cancer Genome Project (PCGP), a three-year, $65 million effort to sequence major pediatric cancers. The latest release of data from the project includes 520 genome sequences from childhood-cancer patients; half the genetic material comes from their tumors and half from their healthy tissues. By matching the tumor genomes to those of normal cells from the same patients, the researchers hope to pinpoint the differences and get a better idea of how the cancer cells got started. The more samples and the more tumor DNA that is layered into the PCGP, the more easily scientists can read the hidden language of cancer cells. While researchers are still a long way from directly applying genomic information to cancer care, a study announced in June 2012 identified potential drug targets in childhood brain tumors. "There is a wealth of information that we are generating, and it will take years to extract all the valuable information," says Dr. James Downing, scientific director and leader of the PCGP at St. Jude Children's Research Hospital in Memphis, which partnered with Washington University in St. Louis for the project.

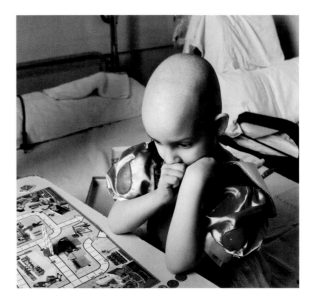

CHILDHOOD-CANCER RESEARCH HAS YIELDED
SIGNIFICANT INSIGHTS
INTO AGGRESSIVE CHILDHOOD CANCERS
OF THE RETINA, BRAIN STEM, AND BLOOD.

IT JUST AIN'T SO ...

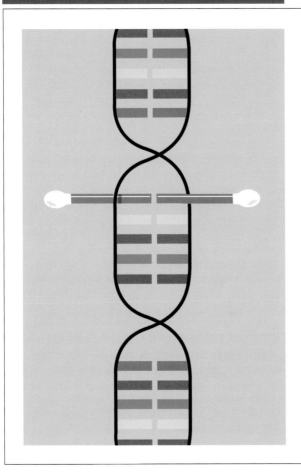

Genes Don't Dictate Your Rx

Our genome makes us who we are; it can provide valuable clues about our health and potentially help us predict our risk for diseases. But a new study shows that knowledge of our DNA isn't as revealing as doctors hoped.

In a report published in *The American Journal of Human Genetics,* scientists at the Harvard School of Public Health found that incorporating genetic information did not improve doctors' ability to predict disease risk above and beyond standard risk factors, including family history, lifestyle, and behavior. So having detailed genetic information didn't change doctors' prevention or treatment plans.

"For most people, your doctor's advice before seeing your genetic test for a particular disease will be exactly the same as after seeing your tests," said Peter Kraft, a coauthor of the research and an epidemiologist at the Harvard School of Public Health, in a statement.

The researchers looked at risk factors—both genetic and environmental—for three common, chronic diseases: breast cancer, type 2 diabetes, and rheumatoid arthritis. All conditions are known to be influenced by a combination of genetic and lifestyle factors. The researchers wanted to determine whether adding information about the interplay of these factors would improve the sensitivity of disease-risk prediction.

It didn't, which means our current limited ability to interpret the complex interplay between genes and environment makes genetic tests too uncertain in guiding health decisions, at least for now.

Physics

LIMITLESS FUSION POWER GETS CLOSER ■ BETTER RADIOCARBON DATING ■ SUPERFAST COMPUTING
QUASICRYSTALS FROM OUTER SPACE ■ FUSILLI-SHAPED RADIO WAVES ■ ESCAPING FROM A BLACK HOLE
QUANTUM TELEPORTATION ■ NOBEL PRIZES FOR PLAYING WITH ATOMS ■ A MAP OF DARK MATTER

Discovering the Higgs Boson

*A particle that eluded scientists for decades
helps explain why our existence is possible.*

By Jeffrey Kluger

If physicists didn't sound so smart, you'd swear they were making half this stuff up. The universe began with a big bang called, well, the Big Bang. It's filled with wormholes and superstrings, dark matter and galactic bubbles, and assembled from little specks of stuff called fermions and leptons, top quarks and charm quarks, all of it glued together by, yes, gluons—and if you claim you understand a bit of it, you're probably lying too.

That's the trouble with particle physics: it exists on a plane that the brain doesn't visit—or at least most brains don't—and wholly defies our intuitive sense of order and reason, of cause and effect, of the very upness and downness of up and down. So we throw up our hands and turn it over to the scientists, and maybe every few years we read a Stephen Hawking book just to keep up appearances.

But when something really big happens, all that can change. In July 2012, as the Internet buzzed with the news that a wonderfully named God particle had been found, as the term "Higgsteria" was trending on Twitter, as scientists around the world opened champagne, the non-physics-speaking joined in, chattering about a thing called a boson and cheering that the standard model had, in the nick of time, been saved. Now, quick, what's the standard model?

There was an odd and merry disconnect between how little most people truly understood the breaking news from the physics world and the celebratory reaction that nonetheless followed it. "Salk Vaccine Works!" we get. "Man Lands on Moon!" we get. Understanding reports that a team of scientists working for the European Organization for Nuclear Research (CERN) had proved the existence of a particle called the Higgs

GOD PARTICLE *Bigger than mere physics, the breakthrough defies the mathematical and brushes up against the spiritual.*

A Puzzle Piece Is Found

Particles acquire mass by slogging through the cosmic Higgs field; how much they get depends on how much resistance they encounter from the field. The Higgs boson, proof that the Higgs field exists, was detected in July 2012.

Particles

Higgs field

FORCE CARRIERS		MATTER			
Higgs boson	**Photon**	**Quark**	**Proton**	**Neutron**	**Electron**
A tiny particle of the Higgs field, it will disintegrate almost as soon as it's created	A particle of light, it's spun out of electro-magnetic fields	A basic building block of matter, it lives in the atom's nucleus	Made of quarks. It used to be considered a fundamental particle	Very similar to a proton, except it carries no electric charge	Orbits the atomic nucleus. We know free electrons as electricity

boson—physics' white whale since it was first postulated in 1964—is a far harder hill to climb.

But the climb is worth it, because the discovery of the Higgs boson helps explain nothing less than why our existence is possible. The particle—named for the Scottish physicist Peter Higgs, one of the small handful of researchers who developed the idea—is the very reason that any mass at all exists in the universe. Energy is easy. But energy and matter are like steam and ice, two different states of the same thing. If you can't ping energetic particles with something—the Higgs boson, as we've now proved—then planets, suns, galaxies, nebulae, moons, comets, dogs, and people don't exist. A cold and soulless

cosmos may not care either way, but we very much do.

"We are nothing but quarks and electrons and a lot of empty space," says physicist Fabiola Gianotti, who headed one of the two experimental teams at CERN that nailed down the discovery using the Large Hadron Collider (LHC), a $10 billion particle accelerator that crashes protons into one another at 99.9999991 percent of the speed of light. "People ask why it is so important to discover the particle that gives mass. But without mass, the universe would not be the way it is."

"My God!" Gianotti exclaimed, jumping up in her chair after she was brought the readouts proving that the Higgs had been found. Maybe it was just an exclamation, but

ONLY 48 YEARS LATER *Scottish physicist Peter Higgs and other researchers first postulated the boson's existence in 1964. On its discovery he said, "It's an incredible thing that it happened in my lifetime."*

with an energy field through which particles must move the way an airplane has to push its way through a stiff headwind. Higgs bosons suffuse the field and are drawn to the particles; more energetic particles attract more bosons, less energetic ones attract fewer. This clustering gives the particles the solidity we associate with matter—and it does something else too. "The Higgs boson has two functions," says Gianotti. "One is to give mass. The other is to prevent the standard model from going bananas."

Bananas, in this case, means the standard model would fall apart. Avoiding that mess was a half-century job, but the pace picked up dramatically in 2010 thanks to work conducted by the LHC and the subsequently shuttered Tevatron collider outside Chicago. In both facilities, physicists didn't study the proton collisions themselves. Instead they focused on the quantum debris, in the form of other particles, that results from them. The goal was to find some that weigh in at 125 GeV (or billion electron volts), the mass predicted for the Higgs.

Lots of bumps appeared in the data at or around that target weight, but the Tevatron was never powerful enough to pin things down firmly, and the LHC, which went to work in 2008, has come fully online only slowly over the years and did not achieve enough propulsive oomph to prove the Higgs case until 2011. Even then, it took trillions of proton crackups to produce enough readings to get to what physicists call the five-sigma level of certainty—and what everyone else calls the eureka moment.

That happened in the spring of 2012. Gianotti's team and another led by physicist Joe Incandela worked separately, and both turned their findings over exclusively to CERN research director Rolf Heuer. Thus, while the two team leaders knew their own work was yielding positive results, only Heuer knew they had both shot bull's-eyes.

"When I saw the first plot from Joe and the first plot from Fabiola, I thought, OK, we have it," says Heuer. "When we all sat down together, I had to spell it out to them. They were reluctant to use the word 'discovery,' but I persuaded them that yes, we can use it."

The announcement of that discovery was made on July 4 to an exuberant crowd of physicists at the International Conference on High Energy Physics in Melbourne. A somewhat dazzled-looking Peter Higgs, now 83, was in attendance and received a long and warm ovation. "It's an incredible thing," he said, "that it happened in my lifetime."

There will never be much return on investment—at least in the traditional sense—in the work at CERN. The field will spin out no Teflon or faster processors or global wireless service the way the space program did. But it is already paying other, far more valuable dividends. The boson found in the deep tunnels at CERN goes to the very essence of everything. And in a manner as primal as the particles themselves, we seemed to grasp that. Despite our fleeting attention span, we stopped for a moment to contemplate something far, far bigger than ourselves. And when that happened, faith and physics—which don't often shake hands—shared an embrace.

the empiricist nonetheless took care to correct herself at the press conference later. "Thanks, nature!" she called out. But it was too late; the cat was out of the bag. She and her colleagues were grappling with something bigger than mere physics, something that defies the mathematical and brushes up—at least fleetingly—against the spiritual.

Despite its bland name, the standard model of particle physics describes some pretty elegant stuff. Completed in the 1970s after decades of work by physicists all over the world, the theory describes three of the great engines that run the universe: the weak nuclear force, the strong force, and electromagnetism.

The weak force is carried by two particles—the W and Z bosons—and, as its name suggests, bonds matter loosely and over very short distances. Its tenuous grip on things is what leads to radioactive decay and, much more happily, initiates the hydrogen fusion that keeps the lights burning in stars like the sun. The strong force is a more robust thing: it causes protons and neutrons to come together in the nucleus of an atom. Carried by particles known as gluons, it is also the force that binds the quarks that make up protons. Electromagnetism is the force behind such phenomena as light, and other everyday waves from radio to X-rays.

Neat, simple, almost intuitive. Except for one thing: all the particles at play in the model—except photons, which transmit light—have mass. And mass needs something to coax it into existence. Enter the Higgs boson. As Higgs and his collaborators explained things, the universe is filled

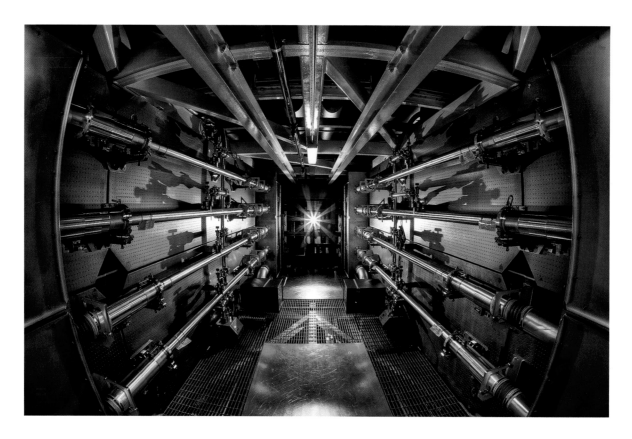

192 LASERS AT A TIME *The experiment's ultimate goal is to create an endless series of tiny explosions that would generate far more power than it took to set them off in the first place.*

THE LASERS DELIVERED

500

TRILLION WATTS OF PEAK POWER—MORE, FOR A FRACTION OF A SECOND, THAN ALL THE ENERGY USED BY THE ENTIRE UNITED STATES AT ANY MOMENT.

A Step Toward Limitless Fusion Power

Think of it as extreme science, a kind of one-upmanship in which researchers fiddle with incredibly complex, painstakingly calibrated machinery to produce unprecedented results—then outdo them.

That's what the National Ignition Facility (NIF) in Livermore, California, did when it pulled the trigger on 192 lasers, all fired within a few trillionths of a second of one another, to deliver 500 trillion watts (500 terawatts) of ultraviolet light—more, for the tiniest fraction of a second, than all the energy used by the entire U.S. at any moment in time.

The NIF's main mission is nuclear-weapons research, but the fusion reactions in H-bombs might also be tamed as source of climate-friendly electricity. The 500-terawatt laser shot was a step in that direction. The ultimate goal is to create an endless series of tiny explosions that would generate far more power than it took to set them off in the first place, just as a stick of dynamite generates more power than the match it took to light the fuse.

Electricity produced by sustained, controlled fusion reactions is a long way from being commercially viable: depending on whom you talk to, it might take 30 to 40 years, and might ultimately prove impossible. But the only way to find out is to try, and the laser blast is a major step forward. So is the construction of a giant fusion machine in France known as ITER, for International Thermonuclear Experimental Reactor. That device will use an entirely different technique—confining a cloud of deuterium and tritium in a doughnut-shaped magnetic "bottle" and using radiofrequency beams to heat it up to millions of degrees, so the atoms' nuclei fuse together and release a stream of energy.

With research moving ahead on two independent paths, we may be getting close to what you might call "the end of the beginning" of a very long and expensive road toward a carbon-free and theoretically limitless source of energy.

How Old Is It? Now We Can Tell More Precisely

Radiocarbon dating has been a workhorse of archaeology for more than half a century, allowing scientists to calculate the ages of bones, ashes, and other organic matter going back nearly 60,000 years. It works because radioactive carbon-14 from the atmosphere, absorbed by living things, begins to decay into nitrogen-14 once they die. The ration of nitrogen to carbon in a given sample is a guide to how long ago the death occurred.

But this timeline has to be calibrated by comparing it with other dating methods, including tree rings and layers of sediment, which can introduce errors of up to hundreds of years. Newly excavated sediments from the bottom of Japan's Lake Suigetsu are tightening things up, however. According to a study published in *Science,* the alternating layers of algae, clay, and other sediments visible in core samples are so pristine that scientists can count off every year going back 52,800 years. The carbon-14 in organic residues from each layer can thus be dated precisely. With the accuracy this adds to the radiocarbon technique, researchers can get a better handle on species extinctions, abrupt climate change, and the spread of modern humans across the prehistoric world.

THE LAKE SEDIMENT
FROM JAPAN IS SO PRISTINE THAT SCIENTISTS CAN COUNT OFF EVERY YEAR GOING BACK 52,800 YEARS.

PREHISTORIC RECORD *A leaf sample dated to 24,700 years.*

Really, Really Fast Computing

Nobody without at least several advanced degrees entirely understands quantum computing, but the Joe Sixpack version goes like this: regular computers think in ones and zeroes, but they do it with such blazing speed that they can calculate a trajectory to Mars, create the quasi-realistic universe of *WALL-E* and *Toy Story,* and make Siri sound only marginally annoying on the iPhone.

For really tough problems like high-level cryptography, however, a quantum computer is far more capable. Its basic unit of calculation is the qubit, which is both a one and a zero at once (don't ask—it's like Erwin Schrödinger's famous quantum cat, which could be both dead and alive at the same time). As a result, Oxford physicist David Deutsch has said, while today's PCs process a single computation at a time, one after another, a quantum computer could be crunching through a million calculations simultaneously.

Qubits have actually been created in the lab, but they tend to work best near absolute zero (-459.67°F). Now, however, a team of Harvard scientists has managed to create qubits out of impurities in diamonds and get them to store information for nearly two seconds at room temperature. That may not sound like much—but it's a time frame that the researchers claim is six orders of magnitude longer than prior attempts.

IRREGULAR PATTERN *No one believed quasicrystals could exist until one was created in the lab (a computer model, above).*

THE METAL ALLOY

WAS FOUND IN A ROCK IN A FLORENCE MUSEUM, WHERE IT HAD BEEN BROUGHT FROM THE KORYAK MOUNTAINS IN SIBERIA. IT MAY BE AS OLD AS THE SOLAR SYSTEM.

Quasicrystals From Outer Space

Until the 1980s, quasicrystals were considered a fantasy. In the real world, everyone knew crystals were made of atoms tied together in endlessly repeating geometric patterns—cubes, triangles, and hexagons, for example, which fit together neatly, with no gaps in between. Pentagons and some other shapes, however, were thought impossible, since they don't fit together in repeating patterns. Mathematicians were able to draw what such crystals might look like, and even created floor tiles with patterns that don't repeat in the traditional way, but nobody believed they could exist in nature.

Then Israeli chemist Dan Shechtman created a quasicrystal in the lab, silencing the skeptics who said they couldn't exist (he won the 2011 Nobel Prize in chemistry for this achievement). In 2008, an Italian mineralogist named Luca Bindi found a tiny grain of natural quasicrystal, an alloy of aluminum, copper, and iron, in a chunk of rock in a Florence museum. And now, after a harrowing three-week expedition into the Koryak Mountains in Siberia, where the rock was originally found, Princeton physicist Paul Steinhardt has deduced that it came from outer space. The rock is clearly extraterrestrial, and may date all the way back to the formation of the solar system.

This discovery's implications are probably huge, but also rather hazy at this point. "What is clear, however," write Steinhardt, Bindi, and their coauthors in *Proceedings of the National Academy of Sciences,* "is that this meteoritic fragment is not ordinary. Resolving the remarkable puzzles posed by this sample will not only further clarify the origin of [quasicrystals] but also shed light on previously unobserved early solar system processes. Fitting all these clues together in a consistent theory of formation and evolution of the meteorite is the subject of an ongoing investigation."

Fusilli-Shaped Radio Waves

The explosion of mobile phones, wireless laptops, and other devices over the past two decades has placed a real strain on the radio spectrum. Put too many channels too close to each other in frequency, and you run the risk of interference that can scramble them all into incoherence. A team of physicists in Italy and Sweden may have found a way out of this dilemma, however: they've coaxed radio waves into twisting as they go, in a pattern reminiscent of the shape of fusilli pasta. By doing so, they explain in the *New Journal of Physics,* they've shown that two independent waves can operate on precisely the same frequency without interference.

They did it by bouncing radio waves off reflectors shaped something like a spiral staircase, which sent them twisting off into space. It worked in the

lab, but the scientists chose to make a bigger splash with a public demonstration in Venice's Piazza San Marco. After an explanatory sound-and-light show projected on the Palazzo Ducale, the scientists write, "when the signal was tuned from vorticity zero to vorticity one at the same frequency and transmitting simultaneously, a rifle shot was heard, in honor of the first radio transmission made by Guglielmo Marconi in 1895. After this, on the facade of Palazzo Ducale the words 'segnale ricevuto', which in Italian means 'signal received', were projected."

How to Escape From a Black Hole

Most people think anything venturing too close to a black hole gets swallowed up, never to be seen again—but scientists know that's not always true. All over the cosmos, giant black holes at the center of galaxies send powerful blasts of superheated gas and dust shooting outward at high speed.

Astronomers have catalogued thousands of these jets over the decades but haven't been able to figure out how matter spiraling down into a black hole suddenly wrests itself free with such titanic force. Now, thanks to a study in the journal *Science,* there appears to be an answer.

What astronomers already understand, thanks to Albert Einstein, is that every black hole is surrounded by an "event horizon"—a place where matter reaches a point of no return in its inward plunge. Material at the event horizon forms a so-called accretion disk, a concentrated swirl of dust and gas that orbits the hole at nearly the speed of light as it heats and spirals inward. It's at that point that something happens to produce the jets.

To find out what, observers focused a network of radio telescopes on a black hole in the M87 galaxy, 54 million light-years from Earth, allowing them to zero in on the black hole's accretion disk with a resolution 2,000 times that of the Hubble telescope. M87's event horizon, the scientists learned, is about the size of our solar system, and the jets appear to emanate from a region about five times larger. They're probably caused by twisting magnetic fields that propel matter outward with enough force to overcome the black hole's hellish gravity.

The astronomers' next step: boosting their telescopes' sensitivity 10-fold to produce detailed images of the process. As good as the high-speed jets are at escaping black holes, avoiding astronomers' prying eyes will—with luck—prove much more difficult.

ZEROING IN *To learn what happens inside a black hole, astronomers focused on a galaxy 54 million light-years from Earth.*

OBSERVATORIES IN ARIZONA, CALIFORNIA, AND HAWAII ACTED AS A SINGLE, MASSIVE INSTRUMENT WITH A RESOLUTION

2,000

TIMES THAT OF THE HUBBLE SPACE TELESCOPE.

MATERIALIZING *Could* Star Trek's *transporter be a reality someday?*

Quantum Teleportation Leaps Ahead

Einstein was a co-inventor of quantum theory, but he reeled at one of its implications: that particles could affect each other instantaneously, even when widely separated. He pooh-poohed this as "spooky action at a distance," and insisted it simply couldn't be possible, no matter what the equations suggested.

But Einstein was wrong for once: experimentalists demonstrated in the 1980s that photons, or particles of light, behaved just this way. Those first tests spanned only a few feet, but an international team of physicists has now shown it's just as true over a much longer distance—89 miles, to be precise. They created a pair of photons whose polarization—the direction in which they vibrate as part of a wave of light—was identical. By a well-known rule of quantum physics, however, that direction was not only unknown before it was measured; it didn't even exist. The scientists sent one photon zinging from La Palma, in the Canary Islands, to its sister island Tenerife. When the particle got there, they measured its polarization, and that measurement immediately fixed the polarization of its twin particle back home.

That's not the teleportation part. But the simultaneous fixing of both photons' polarization allows scientists to decode information sent in a more conventional way. If this seems too complicated to understand…well, it is. But, say the scientists, "our experiment verifies the maturity and applicability of such technologies in real-world scenarios, in particular for future satellite-based quantum teleportation." And that's got to be a good thing, right?

WHAT EINSTEIN CALLED "SPOOKY ACTION AT A DISTANCE" MAY ACTUALLY BE A WAY OF SENDING THINGS VIA
TWIN PARTICLES.

Nobels for Spying on Atoms

Atoms and subatomic particles can behave very strangely when you're not looking at them—jumping from one place to another without crossing the space in between, being in two places at once, acting like particles and waves at the same time, and more. When you try and measure these bizarre properties, however, the tiny objects tend to start acting perfectly normal. "Nothing out of the ordinary going on here," they seem to be saying. "Just move on, please."

The 2012 Nobel Prize in physics went to two scientists who figured out ways to spy on such bits of matter without tipping them off, opening up new methods of studying them in their untamed state. David Wineland of the National Institute of Standards and Technology (formerly the National Bureau of Standards) did it by trapping electrically charged atoms in electromagnetic fields at low temperatures; Serge Haroche of France's École Normale Supérieure focused on light particles, or photons, hemming them in between supercooled mirrors. There's no telling what practical applications might come from the research—and indeed, Wineland and Haroche had none in mind—but given that all modern electronics springs from our current, crude understanding of quantum behavior, the possibilities (the Nobel committee mentioned quantum computers and superaccurate clocks) are more or less limitless.

SHREWD OBSERVER *Nobel-winning physicist David Wineland.*

INVISIBLE *Scientists have created the largest map of dark matter.*

A Cosmic Optical Illusion

While it outweighs the galaxies that astronomers can see in their telescopes, dark matter is utterly invisible. But thanks to a cosmic optical illusion predicted by Einstein, scientists have created the largest map ever made of how dark matter is distributed through space.

The illusion comes because of general relativity's dictum that massive objects literally warp the space-time around them, distorting our view of anything that lies beyond. So astronomers looked at distant galaxies across a wide swath of sky and noted how their shapes were subtly distorted from what you'd normally expect. A single funny-looking galaxy didn't tell them much. It might just be an oddball. Find similar distortions in thousands of galaxies, though, and you can assume that something is warping their images.

So astronomers calculated the distribution of dark matter in the foreground that would create the patterns of galaxy distortion they saw in the background. The answer: dark matter is spread through the cosmos like an immense spider web, with clumps and filaments interspersed with largely empty expanses of space. It's not the first dark-matter map to show such a structure, but it spans more than 1 billion light-years, far more than any of its predecessors.

IF YOU NOTICE
DISTORTIONS IN THOUSANDS OF GALAXIES, YOU CAN ASSUME THAT SOMETHING IS WARPING THEIR IMAGES.

Einstein Was Right: Never Mind That Faster-Than-Light Particle

The universe as we know it has been saved. The instrument of its salvation, and that of the edifice of physics itself? A fiber-optic cable at the European Organization for Nuclear Research (CERN) near Geneva.

The universe was first endangered back in September 2011, when a group of CERN physicists fired a swarm of neutrinos—ghostly particles that don't give a fig about objects in their path—to a receiver beneath Italy's Apennine Mountains, 450 miles away. According to the receiver, the neutrinos made the trek faster than the speed of light, by 60 nanoseconds—and Einstein promised that nothing in the universe can do that.

If the neutrinos really had gone that fast, it would have shaken the foundation of Einstein's special theory of relativity, which is itself the foundation of more than a century of physics, and fundamental to our entire understanding of the universe. Naturally, scientists were a bit concerned.

They were even more concerned when the researchers reported that, yes, they had checked their instruments and calculations and saw no flaws. Physics was in big trouble.

Or maybe not. A few weeks later, investigators announced that a loose cable had scrambled the timing system. When it was fixed, the 60-nanosecond anomaly disappeared. CERN scientist Sergio Bertolucci had never been worried, he said. Asked at the outset of the debacle if he believed the neutrinos really did travel so fast, he said he had his doubts, "because nothing in Italy arrives ahead of time."

NOT SO FAST *Einstein's special theory of relativity is vindicated.*

Medicine

THE HAZARDS OF EXTREME WORKOUTS ■ A CALCIUM TEST FOR HEART RISKS ■ SHARK ATTACK ON VIRUSES
BIRTH-CONTROL PILL FOR MEN? ■ PROTECTION FROM NICOTINE ■ BPA AND OBESITY IN KIDS
AN HIV-PREVENTION DRUG ■ STEM CELLS TO RESTORE HEARING ■ A FLASHLIGHT THAT BLASTS BACTERIA

The Cancer-Screening Dilemma

*How doctors are getting smarter about who should be screened,
and when. (Hint: it's not for everybody.)*

BY ALICE PARK

There's a simple power behind the idea that preventing disease is better than treating it. The wisdom of halting wayward cells before they form tumors, or keeping blood vessels free of dangerous plaque, is not something that requires a medical degree to appreciate.

This is especially true in the case of cancer; scan the body for the smallest grouping of potentially malignant cells, and you can not only protect people from advanced disease but frequently save their lives. Regular screening with mammograms, Pap smears, and the prostate-specific antigen (PSA) test, as well as checks for skin and lung cancers, have long been seen as the ticket to avoiding the worst forms of cancer and a way to help ensure a relatively healthy life. It was hard to see the downside to all this caution.

But there actually is a downside, as the United States Preventive Services Task Force (USPSTF) found in recent years. It turns out that the public-health recommendations that women get routine mammograms, and that men get regular PSA tests, were based more on intuitive common sense than a hard accounting of the harms or benefits of doing so. And what used to be a simple message—get screened to avoid potentially life-threatening disease—has recently gotten a bit more complicated.

In 2009, the USPSTF stunned doctors and patients alike when it recommended that women no longer start yearly mammogram screening for breast cancer at age 40, but wait instead until they reach 50. That's a decade during which, some breast-cancer-patient advocates argued, thousands of women might develop the disease and end up losing their lives because their tumors weren't detected early enough to allow them to begin treatment. Three years later, the task force took the equally dramatic step of advising

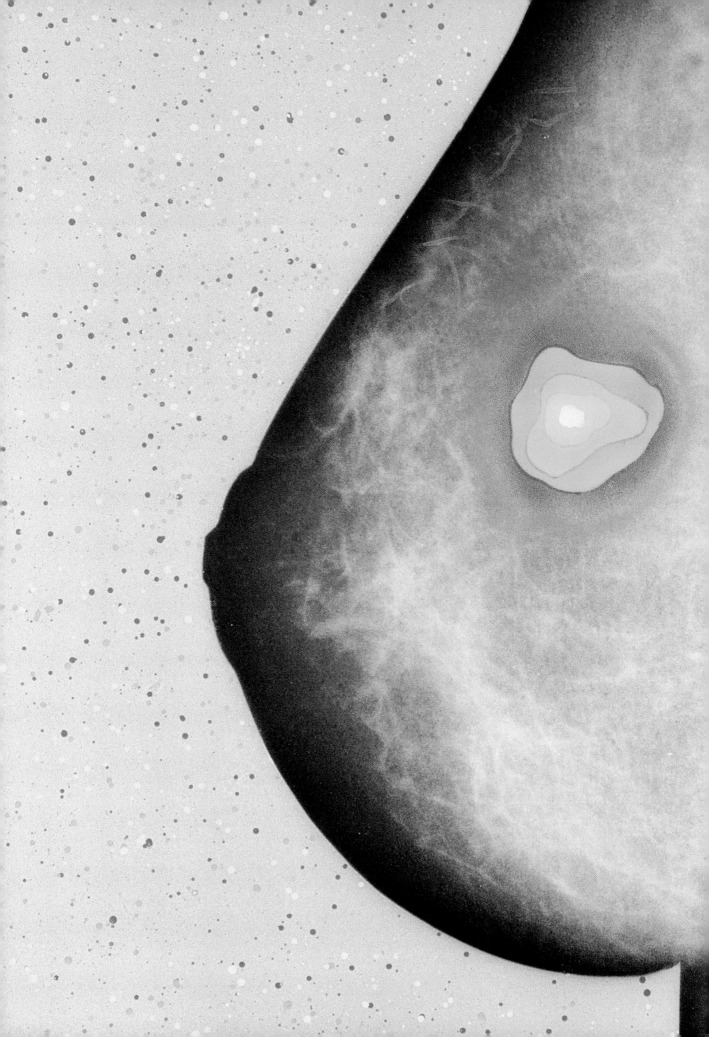

healthy men to stop regular PSA testing altogether.

What changed? A lot, according to the members of the task force, a group of experts authorized by Congress to assess questions of public health in the most rigorous way possible. Many of the studies on which screening guidelines are based don't take a full accounting of the costs of testing, which include not only the number of lives saved from disease that is caught early and treated, but also the additional procedures and complications that arise from false positive results. Take mammograms. For women with dense breast tissue, which includes most women in their early 40s and below, mammograms can be inconclusive or show evidence of disease that isn't there. To help ensure that they aren't missing any potential tumors, physicians may request additional tests, including a biopsy, which comes with its own complications such as infection. If those results turn out to be negative, then the costs of the additional procedures, both in health and in dollars, will not have resulted in any comparable benefit.

When these costs are factored into the equation, it turns out that yearly mammogram screening does not result in enough benefit for younger women compared with the risks it generates. In order to save one life among 40- to 49-year-olds, doctors would have to perform yearly mammograms on 1,904 women over 10 years. That's why the USPSTF advised women to begin screening at age 50, when the benefits start to justify the potential risks of the screening. "I don't always agree with the weighting [the task force] puts on harms, but I do see that the biggest value that the USPSTF brings is an appreciation of the harms associated with cancer screening tests," says Dr. Therese Bevers, medical director of the cancer-prevention center at MD Anderson Cancer Center, headquartered in Houston.

That idea takes some getting used to, especially since the need to catch cancer early is so entrenched in our understanding of the disease. "To most people, cancer means that a little bit turns into a little bit more and then a lot, and a lot [of cancer] spreads and then you die," says Dr. Michael LeFevre, a professor of family and community medicine at the University of Missouri School of Medicine and co–vice chair of the USPSTF. "The notion that cancer doesn't always progress, or doesn't need to be found early, is a pretty foreign concept."

Indeed, when the USPSTF announced its recommendation, breast-cancer-patient advocates immediately criticized the conclusion, concerned that a spike in the number of women with advanced cases of breast cancer, which is much harder to treat and more likely to result in death, would result. In the year following the recommendation, 54,000 fewer women in their 40s received mammograms, a drop of nearly 6 percent, although it's not clear whether that change led to the result the advocates feared.

The task force's case became even stronger when it looked at the evidence for prostate-cancer screening in men. Since 1994 the PSA test, which scans blood for a protein churned out by prostate tumors, has become a mainstay of routine physicals for all men over age 50. Experts acknowledge it's not perfect, and the reason gets to the heart of the

new thinking on cancer screening. Unlike most cancers, prostate cancers tend to be slow-growing, and because tumors tend to emerge later in life, in many cases the cancer may actually outlive the patient. Most men die of other causes—heart disease, for instance—despite the fact that they harbor prostate tumors. To add to the confusion, the PSA test picks up signs of abnormally growing prostate cells, which in some cases might be cancer but in others might be less threatening growths known as benign prostatic hyperplasia.

High PSA readings, then, almost always require some type of follow-up care, which ranges from more frequent PSA screening to biopsies or other procedures to analyze the prostate more closely. After reviewing the data

Constant Vigilance?
Early detection saves lives, but it can also mean misdiagnoses and unnecessary worry.

on prostate-cancer rates, mortality from the disease, and complications resulting from false positive results, the task force found that men who faithfully received yearly PSA screens were no less likely to die from prostate cancer after 10 years than men who did not get the blood test. "These men were living a longer proportion of their lives knowing they have cancer, but they weren't dying at a later date," says Dr. Otis Brawley, chief medical officer of the American Cancer Society.

The USPSTF therefore advised that doctors scrap PSA testing for all men except those with a family history of the disease, or other risk factors that would warrant more rigorous monitoring for tumors.

So why did health experts urge people to get regular mammogram and PSA screening? That advice was based on studies that did not always follow patients for long enough periods of time to assess the impact of regular monitoring on death rates, say the task-force experts. "Now," says LeFevre, "the question isn't just 'Do you do well if you find cancer early,' it's 'Could you do just as well if you found it later.'"

Such an understanding doesn't mean that screening is no longer valid or that it's an unimportant part of cancer care. It simply means that patients and doctors are more informed about the risks and benefits of routine surveillance for tumors. People at high risk of cancer should still take advantage of regular screening, but for healthy individuals without obvious risk factors for the disease, the calculus is changing, and that includes a reassessment of the blind acceptance of the idea that screening can't hurt. "Our mission is to help people understand the pros and cons of screening so they can make decisions," says Dr. Steve Woloshin, professor of medicine at the Dartmouth University Institute for Health Policy and Clinical Practice. "And the better informed people are, the better they will be able to make the right choices about their health.

Cancers that caused the most deaths in 2012

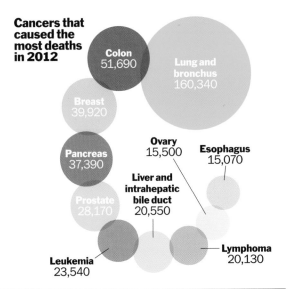

- Colon 51,690
- Lung and bronchus 160,340
- Breast 39,920
- Pancreas 37,390
- Prostate 28,170
- Ovary 15,500
- Esophagus 15,070
- Liver and intrahepatic bile duct 20,550
- Lymphoma 20,130
- Leukemia 23,540

Cancer Celebs

Famous survivors and their relatives have the power to increase screening rates—for better or for worse.

KATIE COURIC
After her husband died of colon cancer, Couric had a colonoscopy on live TV. Use of the procedure spiked 20%.

BOB DOLE
Diagnosed with prostate cancer, Dole had his prostate removed and urged men to get PSA tests.

BETTY FORD AND HAPPY ROCKEFELLER
The first and second ladies admitted having breast cancer, then a taboo topic.

MEGHAN MCCAIN
Her father, Senator John McCain, had melanoma surgery on his face, and she starred in an ad to raise awareness.

Benefits vs. Harm

Saving one life via screening often means unnecessary worry, biopsies, and treatment for many others. Prostate-cancer screening using a blood test called PSA is one example.

PROSTATE CANCER

1,000
men ages 55 to 70 undergoing annual PSA testing for 10 years

1
might avoid prostate-cancer death

150–200
will have a false alarm requiring a biopsy

30–100
will be overdiagnosed and receive an unnecessary radical prostatectomy or radiation

---

Screening's Recent History

Early detection is considered effective if more diagnoses mean fewer deaths. Here's a snapshot of when this is and isn't the case, along with guidelines from a government task force on screening for various cancers.

DIAGNOSES VS. MORTALITY RATE
PER 100,000 AMERICANS

SHOULD YOU GET SCREENED?

DIAGNOSES IN 2012

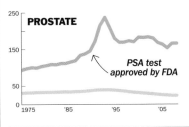

BREAST
Mammograms become widespread

YES Get a mammogram every other year beginning at age 50. Start earlier if you and your doctor decide that's best or if you have risk factors, like a family history of breast cancer.

229,060

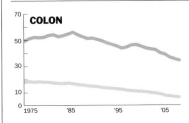

COLON

YES Get a fecal occult blood test, sigmoidoscopy, or colonoscopy at age 50 and at regular intervals until age 75.

103,170

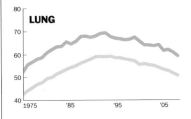

PROSTATE
PSA test approved by FDA

NO Recent evidence seems to indicate that the prostate-specific antigen (PSA) test may do more harm than good.

241,740

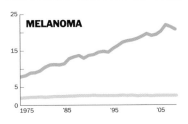

LUNG

MAYBE Two randomized trials to assess lung-cancer screening are under way. For now, the task force says the harm-benefit balance is unclear.

226,160

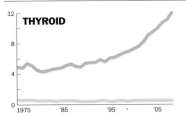

MELANOMA

MAYBE There isn't enough evidence to determine whether whole-body skin examinations by primary-care doctors or patients themselves produce more benefit or harm.

76,250

THYROID

PROBABLY NOT Guidelines are "under review" but aren't a top priority because the mortality rate is so low. A 1996 guideline advised against screening.

56,460

SOURCES, CLOCKWISE FROM TOP LEFT: AMERICAN CANCER SOCIETY; NATIONAL CANCER INSTITUTE; U.S. PREVENTIVE SERVICES TASK FORCE; DR. H. GILBERT WELCH (FIGURES ARE ESTIMATES)

Extreme Workouts: Can Exercise Do More Harm Than Good?

STUDIES SHOW
THAT ENDURANCE
ATHLETES HAVE A

5
TIMES

HIGHER RISK OF
ATRIAL FIBRILLATION,
OR FLUCTUATIONS
IN THE HEARTBEAT
THAT CAN TRIGGER
MORE SERIOUS
HEART PROBLEMS.

Getting at least some exercise is good for us, and more is even better, doctors say. But is there such a thing as too much? Apparently, yes, according to researchers led by Dr. James O'Keefe, a cardiologist at the Mid America Heart Institute of St. Luke's Hospital in Kansas City, Missouri. Reporting in the journal *Mayo Clinic Proceedings,* O'Keefe and his colleagues found that physical activity, like a drug, can be harmful if it's overdone. "More is better up to a certain dose," he says. "But after that, it may detract from health and even longevity."

O'Keefe reviewed studies of people who trained for marathons, triathlons, ultramarathons, or long bike races. Overall, exercise can extend life up to seven years, but that's not the case with extreme athletes. During and immediately following a marathon, runners show up to a 50 percent increase in levels of an enzyme called troponin, which signals damage to the heart (it's the same enzyme that shoots up in patients having heart attacks). In the case of extreme exercise, troponin may start to climb as heart-muscle fibers tear under the burden of pumping continuously at a high level. "When you're sitting around, your heart is pumping about five quarts of blood a minute," says O'Keefe. "If you go and run for 26 miles, or do a full-distance triathlon, the heart is pumping 25 quarts a minute for hours and hours." One result: endurance athletes have a five times higher risk of atrial fibrillation, or dangerous fluctuations in heartbeat, than other people.

Even too much activity is better than none at all, and no matter how much the people in the study ran, they didn't do worse than nonrunners when it came to longevity. The optimal amount of exercise appeared to be about 10 to 15 miles per week. O'Keefe doesn't discourage patients from participating in marathons, but he counsels them to consider the experience a once-in-a-lifetime thing: "If they want to train for a marathon, to cross it off their bucket list, I tell them OK, but it's not a healthy long-term habit."

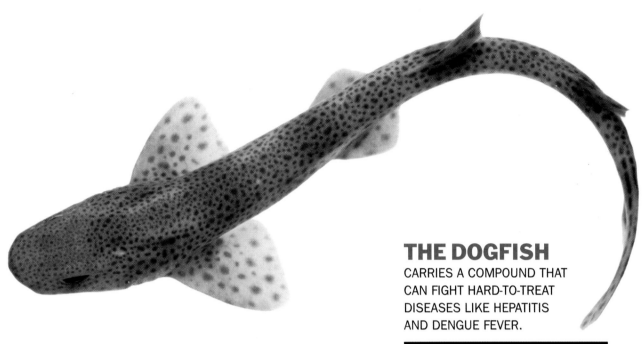

Calcium Test to Predict Heart Attacks

Doctors are already pretty good at predicting heart-disease risk, but a new study shows they could become even better by adding a coronary-artery calcium test. Done using a CT scan, the test assesses the amount of calcium in blood vessels, which can lead to stiffening of the vessels and eventually blood clots. That measure could reclassify about 25 percent of patients more accurately, says Dr. Joseph Yeboah, a cardiologist at Wake Forest Baptist Medical Center in North Carolina. Physicians currently evaluate heart risk using a method that accounts for gender, age, cholesterol level, blood pressure, and smoking status to categorize people into one of three risk groups: low, intermediate, or high. The 28 million Americans in the intermediate category pose a problem for doctors. Should they be treated aggressively with choles-terol-lowering or blood-pressure-controlling drugs? Or would they be better off adjusting their diet and lifestyle? In a study Yeboah conduct-ed of 1,300 patients with intermediate risk, he found that coronary-artery calcium scores can help make that determination, dramatically improving risk prediction. "The question every clinician is asking is which [test] is best," he says. "We put them head-to-head, and the results showed that coronary calcium is better than all the others."

GETTING A CLEARER PICTURE *A partially occluded coronary artery.*

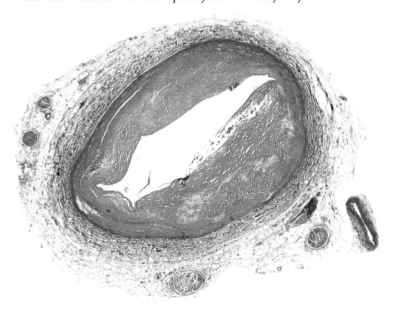

An Antiviral Bottom Feeder

The dogfish, as its name implies, is not the most majestic of sharks. Often considered a trash fish, the four-foot bottom feeder finds itself most popular on dissection slabs in high-school biology class.

But the lowly dogfish emerged as a medical powerhouse when researchers isolated a compound in its tissue that can kill human viruses, including hepatitis and dengue fever. Scientists at Georgetown University believe the substance, called squalamine, works by changing the electrical balance within cells. As a positively charged molecule, it sticks to negatively charged inner cell membranes, then "pops off" other positively charged proteins that viruses need to reproduce. "It looked like no other compound that had been described in any animal or plant before," study researcher Michael Zasloff told LiveScience.

Discovered in 1993, squalamine was first shown to be an effective antibacterial that could treat some cancers and eye disorders. (It's synthesized in the lab, so no sharks are used for research.) The recent discovery could lead to "a whole new approach to the treatment of viral disease," said Zasloff. For now, squalamine works only with blood vessels, capillaries, and the liver—and has shown toxicity in some cell types when given in levels of antiviral efficacy. But human clinical trials are scheduled to begin in about a year, and scientists are hopeful it can be tailored to fight a range of viruses. In the meantime, the dogfish is finally having its day.

THE NOVEL VACCINE

PREVENTS NICOTINE FROM REACHING
RECEPTORS IN THE REWARD CENTERS OF
THE BRAIN, COUNTERING BOTH THE PLEASURE
AND THE ADDICTIVENESS OF SMOKING.

A Smoking Cure—for Mice, at Least

Researchers are getting closer to developing a vaccine that could help protect people from the addictiveness of nicotine. The novel vaccine, which has so far been tested only in animals, prompts the body to manufacture antibodies that are attracted to nicotine. These patrol the bloodstream, soaking up the chemical and preventing it from reaching receptors in the reward centers of the brain. No pleasure means no addiction.

Nicotine vaccines have been tried before but failed because researchers couldn't maintain sufficient antibody levels in smokers' blood. One previous attempt, called NicVax, delivered nicotine encased in a harmless cholera toxin into the body in hopes that the immune system would recognize the invader and make antibodies. But even with the cholera, nicotine is too small a molecule to trigger a robust immune response.

So, rather than delivering nicotine itself, researchers led by Dr. Ronald Crystal, chairman of the department of genetic medicine at Weill Cornell Medical College in New York, tried gene therapy. Crystal and his colleagues used a cold virus to ferry into the body the genes needed to make the nicotine antibody; the vaccine also contained instructions for infecting the liver, which is a factory for churning out proteins and other compounds. Once the vaccine infected liver cells, they began producing copy after copy of the antibody and releasing them into the bloodstream. In mouse experiments, the scientists found that inoculated animals were still making the antibody weeks after receiving the vaccine. And when vaccinated mice were injected with nicotine, the antibodies in their blood bound to it and effectively blocked most of it from entering the brain.

Finally, a Birth-Control Drug for Men?

Aside from condoms and a vasectomy, there aren't any reliable methods of male birth control. But researchers at Dana-Farber Cancer Institute and Baylor College of Medicine have found that an experimental new drug may point the way to a male birth-control pill.

The researchers tested a small molecule they called JQ1. When they injected it into mice, it reduced sperm production to the point that the animals became infertile, while still retaining their sex drive. Over an 18-month period on the drug, the mice produced a normal amount of testosterone and mated as much as usual—just without making any little mice. What's more, when the animals stopped treatment, their fertility was restored in one or two months.

JQ1 works by stealth. It's small enough to cross the blood-testis barrier, so it can reach the cells that make sperm. The molecule inhibits a protein called BRDT, which sperm need to mature. The result is fewer and lower-quality sperm. When treated with an initial, lower dose of the drug, four of seven male mice were still able to re-

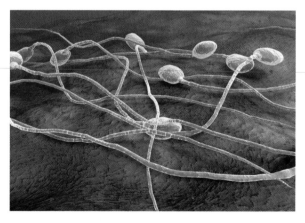

produce, but their litters were smaller than normal. When the dosage was increased, none of the mice were able to sire babies at all. The molecule isn't ready for human testing yet, but since men do have a BRDT gene and JQ1 does appear to target it, researchers hope for a similar result. The birth-control playing field, like so many other things between men and women, may be a lot more level soon.

THOSE IN THE HIGHEST QUARTILE
OF BPA EXPOSURE ALSO HAD THE
HIGHEST RATE OF OBESITY, AT

22%.

THOSE WITH THE LOWEST LEVELS
OF EXPOSURE WERE LEAST
LIKELY TO BE OBESE, AT

10%.

Linked: BPA and Obesity in Kids

Researchers say unhealthy diets and lack of exercise aren't the only culprits in childhood obesity. Bisphenol A (BPA) may share some of the blame. In the latest study involving the endocrine-disrupting plastic additive, scientists found that children with higher BPA levels were likelier to be overweight than kids with lower levels.

In the study, led by Dr. Leonardo Trasande, an associate professor of pediatrics and environmental medicine at New York University School of Medicine, researchers looked at both body-mass index and BPA levels in the urine of 2,838 children aged 6 through 19. They found that those in the highest quartile of BPA exposure had the highest rate of obesity, at 22.3 percent, while those with the lowest levels of the chemical were the least likely to be obese, at 10 percent.

It's possible that instead of BPA contributing to obesity, the heaviest children might simply be consuming more high-calorie foods sold in packaging that contains BPA, such as canned sodas or microwavable products. But Trasande says he and his colleagues corrected for this, finding a strong correlation between BPA and obesity among kids who consumed both more and less than the average number of calories each day. The connection also remained strong after the researchers adjusted for the amount of TV the kids watched—a marker of activity level.

Interestingly, the link was stronger in whites than in any other ethnic group. Since there were no significant differences in the way the various groups ate, the researchers believe there may be some gene-based difference that makes Caucasian populations more vulnerable to BPA. Either way, the chemical isn't good for anyone, and the best solution is for consumers to avoid packaging that contains it whenever possible, and for industry to find and use a substitute—fast.

PBA LEVELS INCREASED

1,221%

AMONG STUDY SUBJECTS WHO CONSUMED
CANNED SOUP FOR FIVE DAYS.

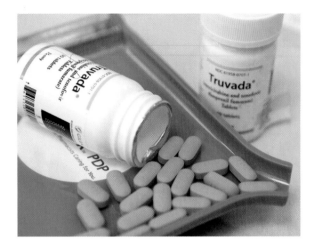

TRUVADA CAN LOWER THE AMOUNT OF HIV IN INFECTED PEOPLE AND, IT APPEARS, HELP SOME AVOID INFECTION ALTOGETHER.

The First HIV-Prevention Drug

Doctors now have another weapon against HIV/AIDS, and it's a potent one. For the first time, the Food and Drug Administration approved a drug treatment that can help prevent infection in healthy people. The drug, called Truvada—which is already approved for the treatment of HIV in infected patients—works by blocking the activity of an enzyme that the virus needs to replicate, thus lowering the amount of HIV circulating in the blood. In clinical trials, the same blocking mechanism was shown to thwart HIV's ability to take hold in healthy cells and start an infection in the first place. The drug does not cure AIDS and does not even guarantee protection from infection, so condoms are still essential, particularly for high-risk people. But one study did show that healthy gay and bisexual men who took Truvada daily and were counseled about safe-sex practices lowered their risk of becoming infected by up to 42 percent. In another study involving heterosexual couples in which one partner was HIV-positive, uninfected partners had a 75 percent lower risk of contracting HIV if they took Truvada.

Stem Cells to Restore Lost Hearing

Here's betting you don't give much thought to deaf gerbils, but be glad that researchers in the United Kingdom do. In a promising early study, they were able to restore hearing in the animals using human embryonic stem cells—an encouraging finding for some of the millions of people who suffer from hearing disorders.

Hearing loss is typically caused by disruptions in the connection between hair cells of the inner ear and the brain. The hair cells turn sounds into electrical signals, which are then carried by auditory neurons to the brain. In some cases those nerves are damaged beyond repair, so replenishing them represents a promising new strategy for restoring hearing.

For the study, the researchers used a chemical called ouabain to damage the auditory nerves of gerbils. Then they injected stem cells that had been nurtured to grow into immature nerve cells into the deafened ears of 18 of the animals; eight received no treatment. Ten weeks later, hearing in the gerbils that received the stem cells had improved by 46 percent, as measured by the brain stem's response to sound. Some gerbils regained nearly full hearing while others regained less, but on average, the animals' hearing was improved to the level at which, in a human, conversation would be possible in an environment with a lot of background noise.

"We have proof of concept that we can use human embryonic stem cells to repair the damaged ear," Marcelo Rivolta, a stem-cell biologist at the University of Sheffield and the lead author of the study, said to Nature News. "More work needs to be done, but now we know it's possible."

Stem cells can be coaxed into any type of cell in the body, and while they have been turned into auditory nerve cells before, this is the first study to show that they can actually restore hearing. "The next goals of any protocol are to gain higher levels of efficiency, reproducibility and safety," Stefan Heller, a researcher from Stanford University who is also working on turning stem cells into hair cells, told Nature News. "Then we can think about patients."

A Flashlight to Blast Bacteria

In a disaster zone, whether it's the aftermath of a tornado or on a battleground, emergency medical personnel have to work fast to save lives. To help them along, scientists have developed a handheld, battery-operated "flashlight" that can instantly sterilize wounds by blasting bacteria with an ionized gas known as plasma, which has become something of a miracle drug in intensive healing. Previous studies showed that it effectively kills bacteria and viruses on skin and water surfaces and helps heal wounds. Why it's such a powerful sterilizer remains unknown, but scientists speculate that reactions between plasma and the surrounding air can create a "cocktail" of reactive molecules containing oxygen, similar to those in our own immune system. Another theory is that ultraviolet radiation plays a role, but the flashlight produces very low UV levels, which makes it safer.

To test the device, researchers from China and Australia created thick biofilms of *Enterococcus faecalis,* a bacteria that frequently infects root canals and is known to be highly resistant to antibiotics and heat. The researchers then exposed some of the biofilms to the plasma flashlight for five minutes and found it was able to kill the bacteria, even at the deepest layer. The real innovation is not just the effectiveness of the plasma treatment but the fact that the flashlight is battery-powered and portable. The device requires further clinical testing before it can be used commercially, but it's getting close, and at about $100 per unit, it could do a lot of healing on the cheap.

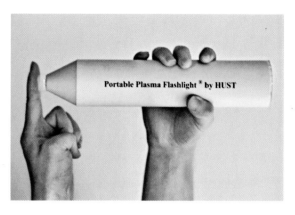

RESEARCHERS CREATED

17 LAYERS

OF BACTERIA AND TREATED EACH WITH A BLAST OF PLASMA, INACTIVATING THE BACTERIA ALTOGETHER.

IT JUST AIN'T SO ...

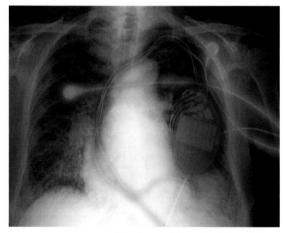

INTERNAL INSECURITY *Your data could be vulnerable to sabotage.*

Medical Devices Aren't Hackproof

A heart defibrillator remotely controlled by a villainous hacker to trigger a fatal heart attack? It may seem like something that happens only in a spy movie. But the Government Accountability Office (GAO) doesn't want to take chances. In a recent report, the agency said the threat that hackers could manipulate defibrillators and other remotely controlled medical devices is real enough for the Food and Drug Administration (FDA) to take action.

In the report, the GAO reviewed research published by security specialists and studies in peer-reviewed journals and determined that these devices are indeed vulnerable to sabotage. In normal use, doctors employ wireless communication systems to download diagnostic status and data on the functioning of such devices and make electronic adjustments remotely. By similar means, control could be tampered with and confidential medical data tapped from the information stream. While no cases of hacking have yet been reported among users, some well-publicized cases of security specialists recently showed it was possible, and alarmingly easy, to hack insulin pumps. That prompted Congress to look into the security issues. "Even the human body is vulnerable to attack from computer hackers," said California Rep. Anna Eshoo.

Fixing the problem isn't easy. Installing security software could put more demand on battery life, for example. And the last thing you want is to require a patient who's having a heart attack to remember a security password before a paramedic or emergency-room team can provide care. The GAO does offer several measures the FDA could implement, including demanding that manufacturers address potential security risks during the pre-market approval process, as well as establishing a separate entity responsible for assessing the security of wireless devices. In a written response to the GAO report, the FDA said it "shares the concern...and emphasizes security as a key element in device design." Patients walking around with one of the gadgets in their bodies would surely agree.

The Earth

HUNTING INVASIVE SPECIES ■ THE END OF THE NORTH POLE? ■ THE DOWNSIDE OF NEW OIL ■ THE BICYCLE BOOM
HELP FOR SMALL FISHERIES ■ THE GRAND CANYON OF ANTARCTICA ■ HAVE WE ENTERED A NEW EPOCH?
THE UPSIDE OF DECLINING BIRTHRATES ■ FORESTS GROWING IN THE ARCTIC

Rainforest for Ransom

*Ecuador's demand: pay us or we'll drill for oil
in the Amazon. Should the world say yes?*

BY BRYAN WALSH

The canoes slip from the dock, the morning mist still clinging to Anangucocha Lake in eastern Ecuador's Yasuni National Park. The Amazon rainforest has yet to fully awaken. Then a small squirrel monkey scurries along a branch arching over the river, followed by another and then another. Soon the trees are full of families of bounding squirrel monkeys. This is wildlife—more vibrant than I've ever seen—treating the great Amazon forest like a playpen. "There are such wonders here," says Luis Garcia, a 47-year-old nature guide and native of the region. "This is why Yasuni is a paradise."

Yasuni National Park—a reserve covering nearly 4,000 square miles on the western fringes of the Amazon basin—is indeed a paradise, considered by many scientists to be the most biodiverse spot on the planet. But it's a paradise in danger of being lost. Oil companies have found rich deposits beneath the park's trees and rivers, nearly 900 million barrels of crude worth billions of dollars. That's money that Ecuador—a small South American country in which a third of the population lives below the poverty line and petroleum already accounts for more than half its export revenue—badly needs, money that oil companies and consumers will be happy to provide if drilling is allowed to go forward. If Ecuador follows the usual path of development, that's what will happen—with disastrous consequences for the park. "Yasuni is a truly unique place in the world," says Gorky Villa, an Ecuadorian botanist who works with the conservation group Finding Species. "Our concern is that it will be ruined before we can even understand it."

But there may be another way. Ecuadorian President Rafael Correa has told the international community that his country would be willing to forgo drilling and leave

A NATURAL WONDER *Yasuni could be the most biodiverse spot on the planet, with an array of plants and animals packed into one section of the Amazon. Clockwise from top left: a tree frog from the Napo River; a yellow-headed calico snake; an orange tiger caterpillar pupating; heliconia flowers; a chestnut-fronted macaw; a katydid molting; a painted-beauty butterfly.*

Yasuni largely intact in exchange for donations equal to $3.6 billion over 13 years, or about half the expected market value of the park's oil deposits. The plan—known as the Yasuni-ITT Initiative—would conserve Yasuni's biodiversity and prevent the emission of more than 800 million tons of carbon dioxide, an amount equal to Germany's annual greenhouse-gas footprint.

The Yasuni plan would be a first for global environmental policy: recognition that the international community has a financial responsibility to help developing nations preserve nature. "Oil is by far the most important part of Ecuador's economy," says Carlos Larrea, a professor at Andean University and a technical adviser on the project. "But we are willing to keep that oil indefinitely unexploited if the international community contributes." Of course, from another perspective, the Yasuni initiative might look like environmental blackmail: pay us or the forest gets it.

There is, however, no ignoring the essential justice of the plan. If we all really do have a shared stake in the natural heritage represented by hotspots like Yasuni, then we have a shared responsibility to help a poor country preserve it. "We need these resources to develop the country, but we're also responsible people who want to protect Yasuni," Correa said in New York in 2011. "If the poor don't receive direct benefits from conservation, conservation won't be sustainable."

In part because it is still relatively uninhabited and undeveloped—canoes are the only way in and out—it's rare to spend more than a few minutes on the creeks that crisscross Yasuni before you catch sight of a fat dragonfly, a rainbow boa, or a golden lion tamarin. "When you go to Yasuni, you will always find new species," says David Romo, an Ecua-

PARADISE UNDER THREAT *Yasuni National Park is still virtually untouched, but the discovery of oil within its boundaries threatens the animals and plants that are found there.*

Ecuador is willing to forgo drilling and leave Yasuni largely intact in exchange for donations equal to $3.6 billion over 13 years, or about half the expected market value of the park's oil deposits.

dorian biologist who has done fieldwork in the park. "It would take us 400 years just to name all the insect species out there." There are estimated to be 100,000 insects per hectare, the highest concentration on earth. More woody-tree species—655 by one count—grow in a single hectare of rainforest in Yasuni than in all of North America. The park is home to 28 threatened or near-threatened vertebrate species—including the white-bellied spider monkey and the giant river otter, which can grow to nearly seven feet—and 95 threatened or near-threatened plant species. It is a bird watcher's paradise, with nearly 600 species, including the white-throated toucan, the phoenixlike hoatzin, and vast swarms of parrots. "The world created a piggy bank of life in Yasuni," says Romo. "The park represents a chance for saving biodiversity in the future—and we have to protect it."

Conservationists fear the effects of oil drilling in and around Yasuni because they've seen the damage that energy exploration can do to nature, and no one knows that better than Ecuadorians. The oil giant Texaco has been accused of polluting vast stretches of the Ecuadorian Amazon with its operations there in the 1970s and '80s, and the company, now owned by Chevron, is involved in a long-running $27 billion lawsuit over the damages, the world's biggest environmental case ever.

If Yasuni is what it is largely because people are absent—with the exception of a few indigenous tribes that live deep within the forest—the amount of industrialization and human activity required to pump 846 million barrels of oil out of the ground would change the park irrevocably. "God gave us the gift of this rich place," says Jiovanny Rivadeneira, general manager of the Napo Wildlife Center, an eco-lodge on the edge of Yasuni. "If there's any oil exploration, we'll feel it first."

If the Yasuni-ITT Initiative moves forward, all of that might be prevented. The plan would require Ecuador to refrain from extracting the oil contained in Yasuni indefinitely in exchange for at least $3.6 billion—half the value of the crude as of 2010—which would go into a capital fund to be administered chiefly by the U.N. Development Programme. That money would be earmarked for investment in renewable-energy projects in Ecuador and social development for indigenous communities in and around Yasuni. As a guarantee, should a future Ecuadorian government decide to go ahead and drill for oil despite the deal, donors would essentially get their money back. "This is the only complete initiative that is out there that is a concrete proposal on how to govern global public goods," says Maria Fernanda Espinosa, Ecuador's minister of coordination of heritage. "The international trust fund is the mechanism for it."

Although Ecuador has so far managed to gather $116 million in commitments from a number of countries and even some individuals, the international community seems unconvinced for the most part. The future of the plan is cloudy at best: though the project met a December 2011 deadline to raise at least $100 million, Ecuador is looking to secure nearly $300 million over the next two years. "We're renouncing an immense sum of money," Correa said in 2011. "For us, the most financially lucrative option is to extract the gasoline."

In reality, the chance of success seems to lessen by the day, but the issues raised by the Yasuni project won't go away. South America is becoming an increasingly important oil producer—the continent holds 20 percent of the world's proven reserves—and much of that crude is buried in and around the Amazon basin. That puts the rainforest in mortal peril: as the global need for oil grows, we're like drug addicts willing to pawn our valuables to pay for the next fix. Yet the burden of protecting our most biodiverse forests—nearly all found in developing nations—can't fall only on poor countries like Ecuador. Each of us benefits from the existence of forest reserves like Yasuni, and each of us should share in the cost of preserving them. If we can't protect the rainforest in Yasuni from the drive for oil, we may not be able to protect it anywhere else.

Why Not Eat the Invaders?

One of the biggest threats to endangered species isn't
hunting or climate change or pollution. It's other animals—
specifically, invasive species that overrun new territory.
With no natural predators, invasive species upend the balance of nature and crowd out natives, causing more than
$100 billion worth of damage each year. Though conser-
vationists spend time and money trying to defend native
species from invasives, too often it's a losing battle.

But there's a force of nature that might be able to check
the threat of invasive species: the human appetite. That's
what hunter and conservationist Jackson Landers argues
in his 2012 book *Eating Aliens: One Man's Adventures Hunting Invasive Animal Species.* Landers tracked, killed, and
ate a number of invasive species, including the Asian carp
in Illinois and the nutria rodent in Louisiana. He has a
great time with his literary stunt, but his book could point
to a new conservation strategy. Humans have managed
to eat species like the bluefin tuna to the point of extinction, after all. Maybe the same impulse could curb the
numbers of invasives. "This is an empirically good value
for hunting," says Landers. "And it doesn't taste that bad."
Armadillo, anyone?

1.
Armadillo
ORIGIN: Central and
South America
INVADED: Texas,
Gulf Coast
THREAT: It eats just about
anything it can get its
snout on, including eggs of
threatened sea turtles.
HOW TO PREPARE: The
meat can be pan-fried
with butter.
TASTES LIKE: Pork

2.
Lionfish
ORIGIN: Pacific and Indian
oceans
INVADED: Caribbean
THREAT: Its venomous
spines are dangerous to
aquatic and human life alike.
HOW TO PREPARE:
Remove the spines—
carefully—and fillet the
fish. Try searing with lemon
and pepper.
TASTES LIKE: Cod

3.
Black Spiny-Tailed
Iguana
ORIGIN: Central America
INVADED: Florida
THREAT: It eats just about
anything, including birds.
HOW TO PREPARE:
Discard the tail and peel the
remaining skin. What's left
makes for good iguana tacos.
TASTES LIKE: Crab crossed
with chicken

4.
Asian Carp
ORIGIN: Asia
INVADED: Missouri and
Mississippi rivers
THREAT: It's a voracious
eater that disrupts the food
chain. It also leaps from
the water at the sound of a
motor, pummeling boaters.
HOW TO PREPARE: Like
whitefish; watch for bones.
TASTES LIKE: Cod

Last Days of the North Pole?

It may sound like a lot: 1.32 million square miles. That's the estimated minimum extent for Arctic sea ice for the summer of 2012, reached on September 16. But it breaks the record for the smallest-ever summer ice cover documented by satellite, beating 2007 by an area equal to the size of Texas. Though Arctic sea ice, which melts and re-forms each summer and winter, usually reaches its lowest extent in mid-September, 2012 actually broke the record even before the end of the melting season. "The strong late-season decline is indicative of how thin the ice cover is," Arctic sea-ice expert Walt Meier said. "Ice has to be quite thin to continue melting away as the sun goes down and fall approaches."

As the summer ended and the long Arctic winter commenced, the sea ice re-formed, but every major melt season like the last few flushes out more of the old, thick multi-year ice, replacing it with thin layers that will melt that much more easily the following summer. If it does, that would continue a long-term melting trend dating back at least until satellite records of the Arctic began to be kept in 1979. The minimum extent in 2012 was 50 percent smaller than the 1979-to-2000 average, which underscores just how much sea ice has collapsed in the past decade.

The question now is whether that trend will continue—and how soon the Arctic will become ice-free. It's a loaded question in climate science, but some researchers worry that that calamity could occur—during the summer, at least—as soon as the end of the decade. That was one projection made by Wieslaw Maslowski, an Arctic expert at the Naval Postgraduate School, who spoke at a Greenpeace panel on the Arctic and climate change in 2011. Regardless of whether Maslowski's timing proves to be accurate, it's clear that the Arctic—the last frontier—is becoming the first frontier of climate change. And as Dan Lashof of the Natural Resources Defense Council said in a statement: "What happens in the Arctic doesn't stay in the Arctic."

SOME RESEARCHERS WORRY THAT THE ARCTIC ITSELF COULD BECOME ICE-FREE— DURING THE SUMMER, AT LEAST—AS SOON AS **THE END OF THE DECADE.**

GREAT WHITE NORTH *Melting Arctic sea ice doesn't add to global sea level (it's already floating), but it's a sign that climate change is accelerating.*

New Oil Edges Out Clean Tech

Oil prices fell throughout much of 2012 after hitting sky-high levels—and that was good news in general for the global economy. Some analysts believe we could even be on the brink of a major surge in global oil production, which would lower prices tremendously. A report published by the Belfer Center for Science and International Affairs predicts that global oil production capacity is likely to grow nearly 20 percent by 2020, to some 110 million barrels a day.

A collapse in the price of oil is bad news, however, for companies working in clean tech, especially transportation technologies. Startups like Massachusetts's A123 Systems, which provides lithium-ion batteries for electric cars, have already gone bankrupt. Plug-in cars like the Chevy Volt have sold more slowly than automakers expected, even with large government subsidies. Cheap oil could set the mainstreaming of electric cars back even further.

This wouldn't be the first time cheap oil undercut what seemed like an imminent green revolution. Alternative-energy investment boomed during the 1970s, when Middle East unrest helped lead to high oil prices and long gas lines. Then when the price of oil collapsed in the 1980s, the nascent alternative-energy sector collapsed along with it. It took years for clean tech to recover from that downturn. With climate change already a dangerous reality—2012 is on track to be the hottest year on record—the world may not be able to withstand another missed chance to break away from oil.

BIG RIG *New discoveries of oil are changing global energy.*

IT'S AS IF JUST ENOUGH AMERICANS HAVE STARTED CYCLING TO PROMPT A BACKLASH—

CALL IT A BIKELASH.

IS THERE ROOM ON THE ROAD FOR EVERYONE?

The Behavioral Factor in a Bike Boom

Once the province of aggro bike messengers and pressed-for-time deliverymen, cycling has gone mainstream in much of New York City. More than twice as many New Yorkers commuted to work by bike in 2011 than in 2006—nearly 19,000—while the number of New Yorkers who ride their bikes daily increased by more than 13 percent over just two years. And much of that two-wheeled growth is due to the surprisingly bike-friendly policies of Mayor Michael Bloomberg and his aggressive transportation commissioner, Janette Sadik-Khan. More than 290 miles of new bike lanes have been built since Bloomberg took over in 2002 (altogether there are more than 700 miles), including new routes physically separated from the streets on main arteries like Manhattan's Ninth Avenue. "Between 2 and 2.5 percent of all vehicle miles traveled in the entire city of New York is by bike," says Charles Komanoff, a New York–based transport analyst. "That's five or six times what it was 30 years ago."

And that is what has some New Yorkers so annoyed. New York is almost certainly the most congested city in America, and even with an extensive public-transit sys-

tem, more than 600,000 cars crawl into lower Manhattan each day. To the drivers and walkers who have long owned the city, any competition is unwelcome. Drivers are frustrated by the unpredictability of cyclists, while pedestrians trade stories about getting buzzed by bikers who ride on the sidewalk or blast through red lights. A 2011 study found that more than 500 New York pedestrians a year make hospital trips after getting hit by bikes.

But look at the numbers more closely, and you'll see that cyclists are far less threatening than their reputation, even in New York. The number of people who've been killed in traffic accidents in New York has declined over the past decade, and the number of bikers is growing faster than the number of bike accidents. And despite the public perceptions of out-of-control cyclists, bikers are rarely at fault for accidents, especially with cars. A study by Monash University in Australia that looked at driver-cyclist collisions found that nearly 90 percent of cyclists had been traveling in a safe manner before the crashes, while vehicle drivers were at fault for more than 80 percent of the collisions, with the remaining collisions classified as no-fault.

Of course, it's also possible that New York's policies could well make the streets safer for cyclists, simply by putting more of them on the streets. In cycling havens like Amsterdam, where 26 percent of daily trips are by bike, cyclists are so commonplace that drivers and pedestrians become accustomed to their presence and give them the necessary space. (It also helps that Dutch children are taught safe cycling from an early age.) The more cyclists there are on the roads, the more they become part of traffic, instead of a resented exception. That in turn helps encourage the cyclists themselves to follow the rules, instead of acting like two-wheeled rebels. If drivers, pedestrians, and cyclists can all remain on the right side of the law, there should be room for all of them.

MORE THAN TWICE AS MANY NEW YORKERS COMMUTED TO WORK BY BIKE IN 2011 AS IN 2006—NEARLY

19,000.

CATCH OF THE DAY *Small fisheries are being depleted, but the right policies can save them.*

Smarter Rules Could Help Small Fisheries

In a major paper published in *Science* in 2012, a team of researchers managed to assess the status of small fisheries, which together make up more than 80 percent of the global catch. What it discovered offers plenty of reason for worry—and some hope as well. It turns out that small, unregulated fisheries are doing worse than the larger fisheries. That means the world's oceans are even more overfished than we had feared.

But the study's authors also found that those fisheries—if better managed—can be rebuilt quickly, and may be able to produce an even larger, more sustainable catch than they do now. One simple way is to provide more funds for scientists to assess smaller fisheries. "The silver lining is that we have proven solutions," said Michael Arbuckle, a senior fisheries specialist with the World Bank.

The *Science* study estimated that 64 percent of the small, unassessed fisheries could "provide increased sustainable harvest" if they came under better management. And that in turn could increase global fish abundance by 56 percent, which would mean more seafood for the world. Given that global food demand is expected to double by the middle of the century, that could be quite a catch.

The Grand Canyon of Antarctica

A team of British researchers has discovered a rift in the rock of West Antarctica that runs as deep as the Grand Canyon. The Ferrigno Rift, as it's called, is bringing more warm seawater into the interior of the Antarctic ice sheet, which can hasten melt that would raise sea level. "The areas that are most vulnerable [to ice melt] coincide with the areas of ancient rifting," Robert Bingham, the discoverer of the Ferrigno Rift and a glaciologist at Scotland's University of Aberdeen, told NBC News. The rifting "preconfigures the topography to a shape that encourages ice loss." In other words, the Grand Canyon of the Antarctic is setting the stage for even faster ice loss than would happen otherwise.

The research expedition was meant to measure the topography of the rock beneath the ice—important scientific work but hardly groundbreaking. Just a few days into the trip, however, radar equipment found that the bed of ice was becoming increasingly thick, indicating a canyon below. And it was a massive one—though unlike the Grand Canyon, which is shaped by erosion, the Ferrigno Rift was created by the forces of continental rifting, the fissures running deep in the earth.

Bingham and his colleagues reported on the rift in a paper published in *Nature* in 2012. But while the research is clearly a victory for topographical knowledge, it matters more for sea-level rise. As Bingham and his coauthors write, the rift's "existence profoundly affects ice loss." The rift allows more warm seawater to get between the Antarctic bedrock and the ice that lies on top of it. The seawater acts as a lubricant, allowing the ice to flow faster into the sea. Glaciers spitting out ice into the sea is a natural process—that's how we get icebergs—but the rate has an impact on the ice cap's melting.

ICE AWAY *Melting in Antarctica is already adding 10 percent to global sea-level rise.*

❝ IT'S NO LONGER US AGAINST 'NATURE.' INSTEAD, IT'S **WE WHO DECIDE** WHAT NATURE IS AND WHAT IT WILL BE."
—*Nobel-winning chemist Paul Crutzen*

Are We Entering a New Geologic Epoch?

For a species that has been around for less than 1 percent of 1 percent of the earth's 4.5 billion-year history, *Homo sapiens* has certainly put its stamp on the place. Humans have had a direct impact on more than three quarters of the ice-free land on earth. Almost 90 percent of the world's plant activity now takes place in ecosystems where people play a significant role. We've stripped the original forests from much of North America and Europe and helped push countless species into extinction. And all the CO_2 that the 7 billion-plus humans emit is rapidly changing the climate—and altering the nature of the planet.

Human activity now shapes the earth more than any other independent geologic or climatic factor. Our impact on the planet's surface and atmosphere has become so powerful that scientists are considering changing the way we measure geologic time. Right now we're officially living in the Holocene epoch, a particularly pleasant period that started when the last ice age ended 12,000 years ago. But some scientists argue that we've broken into a new epoch that they call the Anthropocene: the age of man. "Human dominance of biological, chemical and geological processes on Earth is already an undeniable reality," writes Paul Crutzen, the Nobel Prize–winning atmospheric chemist who first popularized the term Anthropocene. "It's no longer us against 'Nature.' Instead, it's we who decide what nature is and what it will be."

That's why the Anthropocene demands a dramatic change for environmentalism. The reality is that in this epoch there may simply be no room for nature, at least not nature as we've known and celebrated it—something separate from human beings, something pristine. There's no getting back to the Garden, assuming it ever existed. For environmentalists, that will mean changing strategies, finding methods of conservation that are more people-friendly and that allow wildlife to coexist with human development. It means, if not embracing the human influence on the planet, at least accepting it. "We are as gods," writes the environmentalist and futurist Stewart Brand. "And we have to get good at it."

The Good Thing About Low Birthrates

In many ways, the Japanese seem to do everything first: camera phones, Zen Buddhism, small fuel-efficient cars, huge public debt, a stagnant economy, the literary acceptance of comic books—what happens first in Japan eventually makes its way to the rest of the world. And that includes declining birthrates. One of the biggest social issues in Japan is the falling rate of marriage among young people and the vanishingly small birthrate, which translates to an aging—and eventually shrinking—population. But the public angst has made little difference. Most young Japanese women simply don't seem interested in having many children—at least not under the conditions of Japanese society.

Now what began in Japan is happening globally. Fertility is on the decline in much of the world, from Iran (1.7 births per woman) to Russia, where low fertility combined with high death rates mean the population is already shrinking. To some, that means the world is facing what the writer Phillip Longman has called the "gray tsunami"—a moment when the population over 60 years old swamps those under 30. And that includes the U.S., which has long had higher birthrates than most other developed nations.

This would seem to be a slow-motion disaster. Aging countries will face the burden of caring for large elderly populations without a large resource of young workers to draw on. But here's the thing: a Centrum Silver world may have a silver lining for the planet. While overpopulation isn't the human catastrophe it was made out to be in the 1970s, when it seemed as if we were just a few people away from eating Soylent Green, the number of people on the planet—and the amount of stuff they use—is the basic multiplier for nearly all environmental woes, from deforestation to climate change.

The environment—wildlife, plants, the climate—is ultimately the real victim of overpopulation. So maybe a world that grows slower, and grows older, will put less pressure on the environment and buy us a few more years to ensure that our energy use, along with our birthrates, reaches a sustainable level. That will be easier to do if the planet doesn't end up overpopulated. After all, we're supposed to get smarter as we get older. Let's hope that holds true for the species as well.

NEARLY EVERYWHERE AROUND THE WORLD
FERTILITY IS DECLINING,
OFTEN QUITE RAPIDLY, AS WOMEN
BECOME RICHER AND MORE EDUCATED
AND AS THEY MOVE TO CITIES.

Why Trees Take Root in the Tundra

Forests are growing in the Arctic—and global warming is likely the culprit. Researchers in Britain and Finland studied an area of 38,600 square miles in what's known as the northwestern Eurasian tundra, which stretches from western Siberia to Finland. Surveys of vegetation in the region using both satellite data and local observations from reindeer herders showed that in 8 to 15 percent of the territory, willow and alder shrubs had grown into trees more than 6.5 feet tall over the past 30 to 40 years. As Andrew Revkin of the blog Dot Earth puts it, warming has led to "pop-up forests" in regions of the planet that usually see little more than summer shrubs.

That's a sign of just how fast the Arctic in particular can respond to global environmental change. And as the Arctic greens, it could speed warming even more as the darker foliage absorbs sunlight that would have been reflected back into space by the white tundra. While short shrubs can be covered completely in snowfall—thus reflecting sunlight—tall trees are usually above the layer of white.

The advance of forest into the Arctic could increase Arctic warming by as much as 1 to 2 degrees Celsius by

GREEN ARCTIC *Climate change has led to "pop-up forests."*

the end of the 21st century. In a statement, Marc Macias-Fauria of Oxford University, the lead author of the paper, noted how unusual the advance of Arctic forest was: "It's a big surprise that these plants are reacting in this way." The planet is changing, and the Arctic is a bellwether of that change.

IT JUST AIN'T SO ...

You Think Renewable Energy Can't Scale Up? Think Again

Sure, wind and solar and other forms of renewable energy are great for the earth—but they're not really a business, right? The rap against alternative energy has always been that it might work on a small scale but that it can't grow fast enough to displace coal, oil, and other fossil fuels.

But the last few years have put the lie to that argument. Powered in part by ambitious stimulus spending, renewable-energy generation has risen by more than 78 percent since the beginning of 2009, with wind and solar growing the most. Wind accounts for 2.3 percent of the U.S. electricity supply, but in some states—like Iowa and Minnesota—it's already a backbone of the power grid, supplying more than 10 percent of electricity. And the growth in wind is nothing compared with the explosion in solar in the U.S., which is up more than 285 percent since 2009. Thanks to subsidies and advances in manufacturing, the retail cost of solar fell by half during 2011.

The story is even greener in other countries. The share of electricity produced from renewable sources in Germany increased from 6.3 percent in 2000 to 25 percent in 2012. Denmark has long produced more than 20 percent of its electricity from wind, and the country aims to phase out all fossil fuels by midcentury.

It's true that renewable power faces a long climb to displace fossil fuels—and new discoveries of oil and natural gas will make that shift even more economically difficult. But the technology to green the grid is on its way.

Zoology

A CLUE TO THE ZEBRA'S STRIPES ■ ORCAS AND THEIR MOMS ■ THE EMPATHY OF RATS
ZEBRAFISH AS WARNING LIGHTS ■ SCANDAL AT THE HORSE TRACK ■ A RESCUE TEAM OF DOG AND ROBOT
THE DINGO REALLY DID IT ■ HOW THE ALBATROSS SOARS ■ DISCOVERING THE LITTLEST FROG OF ALL

Friends
With Benefits

*Humans aren't the only species capable of forging true
and lasting friendships–and getting many of the same rewards.*

By Carl Zimmer

Since 1995, John Mitani, a primatologist at the University of Michigan, has been going
to Uganda to study 160 chimpanzees that live in the forests of Kibale National Park.
Seventeen years is a long time to spend watching wild animals, and after a while it's rare
to see truly new behavior. That's why Mitani loves to tell the tale of a pair of older males
in the Kibale group whom the researchers named Hare and Ellington.

Hare and Ellington weren't related, yet when they went on hunting trips with other
males, they'd share prey with each other rather than compete for it. If Ellington reached
out a hand, Hare would give him a piece of meat. If one of them got into a fight, the other
would back him up. Hare and Ellington would spend entire days traveling through the
forest together. Sometimes they'd be side by side. Other times they'd be 100 yards apart,
staying in touch through the foliage with loud, hooting calls. "They'd always be yakking
at each other," says Mitani.

Their friendship—for that's what Mitani calls it—lasted until Ellington's death in
2002. What happened next was striking and sad. For all the years Mitani had followed
him, Hare had been a sociable, high-ranking ape. But when Ellington died, Hare went
through a sudden change. "He dropped out," says Mitani. "He just didn't want to be with
anybody for several weeks. He seemed to go into mourning."

For evolutionary biologists and anthropologists, friendship has been considered one
of the core traits of only one species of ape: us. The conventional thinking held that, along
with our capacity to feel love, loyalty, and compassion, our ability to forge long-term,
meaningful bonds with friends set us apart. To the degree that nonhuman animals have

exhibited such traits, they're really just making a genetic calculation. They'll protect family members, but only because they share so many genes. They'll help an unrelated member of their species too, but that's an even colder transaction known as reciprocal altruism: I'll do you a favor today, but I expect one in return tomorrow.

Humans do this kind of interpersonal ledger balancing as well. It's not for nothing that if a friend lends you $10, you feel a faint sense of unease until you pay it back. If we didn't all feel that, *Homo sapiens* would not have become as cooperative a species as it is. But reciprocal altruism is to friendship as reproduction is to romance. In both cases, we start with a primal impulse and then embroider deep feeling into it. Animals, we've always told ourselves, do nothing of the kind.

Mitani and his colleagues now know better. Unrelated chimpanzees, for example, can develop strong bonds that last for years, and long-term studies by other researchers have revealed durable friendships beyond the chimp species. Dolphins make friends with unrelated dolphins, hyenas make friends with hyenas, and the same is true for elephants, baboons, and horses. Not all animal friends exhibit all those behaviors, but they exhibit enough of them—with enough consistency—that something deep is clearly going on.

However widespread animal friendship is, it is changing our assumptions about how nonhuman societies work. It could also change the way we think about our own friendships—and even about our health. It's well established that having close friends can contribute to a longer life and a lower incidence of disease, but it's never been easy to establish why. Studies of animals might provide some answers. Even before that work is done, though, one thing is clear: humans have always known it's hard to get through life without friends, and it appears that animals are wise to that secret too.

In the field of animal-friendship research, charismatic critters like dolphins and chimpanzees get a lot of the attention, but it's baboons—far more distantly related to us than the great apes—that have provided some of the most powerful insights. In the late 1990s, UCLA anthropologist Joan Silk was working with Princeton primatologist Jean Altman on a long-term study of savanna baboons in Kenya's Amboseli National Park. Silk came up with a painstaking method for measuring the strength of the relationships between primates. She and her colleagues went back through their records and randomly selected hundreds of observations of each female baboon from years of fieldwork. Then they determined how often that baboon was sociable—sitting near another individual or grooming it, say—and noted which baboon was pairing off with which.

When the scientists crunched the data, they discovered a complex social world they hadn't noticed before. "They have very strong relationships with some females and weak relationships with others," says Silk. In many cases the strongest bonds were between unrelated females, and those lasted years. To describe these relationships, Silk, who arrived at the work as a skeptic of the whole idea of animal

If having friends leads to having more babies, the friendliness trait gets passed on, becoming common across the species.

friendship, at last began to use what she calls the F word.

Other scientists conducting long-term studies of species noticed something similar going on. In 1970, Randall Wells, now a biologist with the Chicago Zoological Society, began following bottlenose dolphins in Sarasota Bay in Florida, getting to know them so well that eventually he could distinguish one from another simply by the appearance of its dorsal fin. Over time he discovered that some unrelated male dolphins spend considerable amounts of time together in pairs. "Usually they're swimming side by side," says Wells. "The rest of the time we'll see [them] alone, but they'll be back together again within a few hours."

Unrelated female dolphins do things differently. They spend time together during their fertile years, but these bonds are fluid, with individuals moving from one group to another in the bay. Only when they're in their 50s and no longer reproducing do females develop enduring bonds, and those are with just one or two female friends.

One day in 2008, for example, Wells and his colleagues noticed that a 58-year-old female he named Nicklo had swum into the sea-grass meadows next to the lab to hunt schools of mullet. As mullet try to escape, the dolphin whacks them with its powerful tail. A good fish whacking can leave a mullet stunned so the dolphin can make an easy meal of it. But that day Nicklo was not whacking fish on her own. She was on the hunt with an unrelated old female named Black Tip Double Dip. The pair of dolphins drove the mullet schools from different sides, each whacking fish into the air. Wells had rarely seen two female dolphins whacking fish together, but he began to see Nicklo and Black Tip Double Dip doing it more and more often. Sometimes they'd be joined by another old female named Squiggy. So much teamwork, of course, could simply be the utilitarian business of cooperative hunting: if three dolphins work together, all three eat better. But Wells and his colleagues would also find the trio just swimming in tight formation, apparently keeping one another company. It's not quite *The Golden Girls,* but it's not all that different either.

As evidence for the F word piled up, the question shifted from "Do animals make friends?" to "Why do they bother?" The most obvious answer is that friendships boost reproductive odds. If having friends somehow leads to having more babies, the friendliness trait gets passed on, becoming more common across the species. For male dolphins, the reproductive benefit may come from a friend's playing wingman. A single male may have a hard time driving off other males while mating, but two males working together may be able to do the job. Females lean on one another more after their babies are born. A group of dolphin moms will often form circles around their calves, perhaps protecting them from predators. "We call them playpens," Wells says.

In New Zealand, Elissa Cameron of the University of Tasmania studies a population of 400 feral horses in the Kaimanawa Mountains. The horses live in bands that are typically not made up of close relatives. After collecting four years of data, she found that pairs of mares would establish strong bonds, and those bonds endured throughout her study. Cameron compared the strength of a mare's friendships to her reproductive success and discovered that the more close friends a mare had, the more foals she could rear.

One of the most provocative implications of these studies is that friendships that evolved within species may sometimes reach across the species barrier. In her bestselling book *Unlikely Friendships,* journalist Jennifer Holland describes many such surprising pairs—a gorilla and a kitten, a cheetah and a dog, a hamster and a snake. YouTube, a decidedly more ad hoc source, is filled with clips of cross-species buddies.

But what you see on screen may be less authentic than it seems. Barbara King, an anthropologist at the College of William and Mary and the author of *Being With Animals,* thinks a lot of these cases reflect wishful thinking more than actual friendships. "Right now the label is being applied far too broadly and uncritically," she says.

Studies of animal friendships may deepen our understanding of how complex the nonhuman world is, but there are more tangible lessons as well. The better we understand how friendships change an animal's physiology—improving its health in the process—the more we can learn about the power of those processes in ourselves.

If humans came late to the idea that other animals have the same capacity to form friendships that we do and derive the same benefits, it may be that we weren't paying attention. Chimpanzees and baboons share an ancestor with humans, one that lived 30 million years ago. Maybe that monkeylike progenitor formed friendships with its troopmates, and maybe it inherited the ability from a still more distant mammalian grandparent. Even as we all diverged into multiple species, pursuing our very different evolutionary arcs, all of us—Nicklo the dolphin and Hare the chimpanzee and Bob, the guy who's been your best friend since high school—may have retained the simple but powerful ability to find one another and care about one another.

Different Species Experience Friendship in Different Ways

Dolphins
Males form friendships when they're young and maintain them throughout life. Females have fluid friendships when they're in their fertile years, but moms cooperate to protect young. Older females stick to one or two close friends.

Rhesus monkeys
Members of wild troops form close social bonds with a few other members. This results in lower levels of the stress hormones known as glucocorticoids; the reduction can improve health.

Horses
Individuals in the same band are usually not relatives. They pick a few friends and groom or play with them or simply rest their heads against one another's. Heart rates go down during these quiet moments.

Chimpanzees
Researchers have observed chimps sharing meat, coming to each other's aid in a fight, and traveling through the forest side by side. When one dies, the other appears to grieve.

Baboons
Females form close relationships with a select few other females that last for years. Those that have friendships are four times as likely to survive to age 15 as those that don't.

The Zebra's No-Fly Zone

Evading colorblind lions? Creating air turbulence to keep cool? Acting like nature's bar codes so they can tell each other apart? There have been many colorful theories about the age-old question of how the zebras got their stripes. But new research offers a more surprising, if prosaic, explanation: they discombobulate the eyesight of disease-carrying horseflies.

Female horseflies—they're the ones that suck blood—are more attracted to dark animals because of the way they reflect light. Dark fur causes light to vibrate horizontally, much the same as if it were bouncing off water. Flies go for this kind of polarized light, which humans see as glare, because it's a road sign for water, where flies lay eggs and mate. It also directs them to big, dark animals, where they can hover and chomp away. The bites are bad news for zebras—they're not only painful, but they expose them to lethal diseases and distract them from grazing, which can have poor consequences for their offspring.

Zebra embryos start out with dark skin and develop white stripes before birth. To see which kind of animal hides tend to attract the most flies, researchers from Sweden and Hungary went to a horsefly-infested farm outside Budapest, where they painted life-size zebra models all white, all black, or with stripes of various widths. Oil and glue were applied to models to capture the attracted bugs. Researchers assumed the number of insects drawn to the bicolored models would fall somewhere between the all-white and all-black ones. To their surprise, they discovered that zebras stripes were the best fly repellent, and that skinnier stripes were even better, according to the report in *The Journal of Experimental Biology*. This may explain why they appear on zebras' faces and legs, where their skin is thinnest.

The black-and-white pattern reflects different polarizations of light, effectively scrambling the wave and making the zebra more difficult to single out in its surroundings. Scientists don't preclude other factors from influencing the evolution of zebra pigmentation. Still, could the application of stripes be used to help other animals, including us, keep flies from biting?

VERTICAL STRIPES

EACH REFLECT DIFFERENT POLARIZATIONS OF LIGHT, WHICH MAY WARD OFF HORSEFLIES LOOKING FOR A SMOOTH, HORIZONTALLY POLARIZED SIGNAL, WHICH INDICATES THE PRESENCE OF WATER.

Orcas Are Mamas' Boys

It may sound coldhearted to consider, but evolutionarily speaking it's odd that orca moms continue to live for decades after they've finished reproducing. As only one of three species that go through menopause (the others being pilot whales and humans), female killer whales can live into their 90s, which is 50 or 60 years after their genes have been passed to a new generation.

It turns out they're sticking around to help their adult sons, who, despite the reputation as ferocious predators that earned them the nickname "killer whales," live longer when Mom's got their back. Researchers

amassed a four-decade census of 589 Pacific Northwest whales and found that, for males over 30, the death of a reproductive mother meant an eightfold increase in the likelihood of their own death within a year—and the mortality risk increased to 14-fold if the mother had gone through menopause. Clearly, postmenopausal mothers are great to have around, and older sons seem to benefit more than the youngsters.

Scientists believe the mothers increase the transmission of their genes by helping their sons survive. But just how does an old-lady orca help her virile son? Killer whales are very difficult to study in the wild, but speculation includes help with foraging and support during aggressive encounters. Whatever the reason, orca matriarchs and their mamas' boys may help unlock the evolutionary puzzle of menopause.

THE FREE RATS IMMEDIATELY
LIBERATED
THEIR TRAPPED PARTNERS ONCE THEY FIGURED OUT HOW TO OPEN THE RESTRAINT.

Rats May Not Be, Well, Such Rats After All

In the first study of its kind, rats have shown empathy-driven behavior, helping to free a trapped cagemate for no reward other than relieving its distress—even when a stash of delicious chocolate chips is on the line.

Although previous studies have shown that empathy isn't just the province of humans, this is the first evidence that rodents display such behavior. In the study, a pair of rats were put in a testing area where one roamed free while the other was trapped in a plastic restraining tube. The free rat could liberate the trapped one by figuring out how to tip open the door. After a week of determined effort, the Good Samaritan rat learned to open the restraint and liberate its trapped companion.

In an ingenious twist, researchers added chocolate chips to the mix. Rats tend to love chocolate. Would the rats go for the sweets and leave their friends locked up, or open the door and share the goodies?

About half the time, rats chose to free their cagemates and share—a result that surprised researchers. The experiments suggest that empathy may be part of our biological inheritance—a lesson some *Homo sapiens* could learn from our rodent cousins.

How the Zebrafish Illuminates Pollutants

It may look like a party trick, but a fluorescent zebrafish about an inch and a half long is shedding light on the harm pollutants can cause to human and animal health. Genetically engineered by a team of British scientists, the fish glow green in the presence of endocrine disrupters, which are chemicals that alter hormone signaling in the body. Two prime culprits are bisphenol A, or BPA, a synthetic chemical found in plastics used in many everyday products, and ethinyloestradiol, used in contraceptive pills. Excessive human exposure has been linked to breast and testicular cancer, decreased sperm count, and other reproductive problems. The disrupters have also been shown to cause male fish to switch genders. The chemicals end up in rivers, where they find their way to aquatic flora and fauna—and, eventually, humans.

Scientists have had difficulty understanding how these toxins act inside the body. But after exposing the fish to pollutants in different concentrations, they were able to track in real time which specific tissues and organs lit up. The liver, ovaries, and testes had already been identified as targets, but now previously unidentified parts of the body glowed, including the skeletal muscles and the eyes.

We still have many gaps to fill in our knowledge of how environmental estrogens work their way through our bodies, but the zebrafish is one of the most comprehensive and intelligent sensor systems to date for tracking the impact of the chemicals on precise areas of the body, which means a lot for future health assessments. That's a pretty big achievement for a glow-in-the-dark fish.

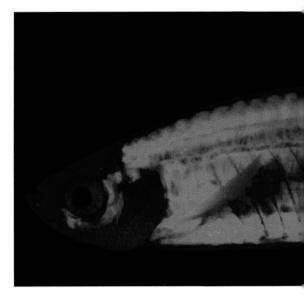

OBSERVING THE

GLOWING FISH

CAN HELP DETERMINE THE THRESHOLDS AT WHICH THE TOXINS AFFECT DIFFERENT ORGANS AND TISSUES.

Uncovering the Frog-Juice Scandal

More than 30 racehorses tested positive for an illegal performance-enhancing substance that rivals the slimiest doping scandal of human athletes, according to *The New York Times*. Racing officials are seriously concerned about increased trafficking in dermorphin, a drug extracted from the back of a South American frog.

Dermorphin has the twofold benefit of numbing the pain a horse might feel from an injury and simultaneously rendering the animal hyperactive. It is a far more powerful painkiller than morphine. This is only one of many doping methods that have plagued racetracks over the years, though it's one of the most exotic, along with cobra venom, which has been used as a pain suppressant.

Rumors about "frog juice" had been circulating for years, but officials had been unable to test for the substance in horses until a lab in Denver recently found the correct procedure. Though regulators suspended a champion trainer for 10 years for using the substance, cheating will abide. Racing has a long history of doping, and officials bemoan the fact that as soon as they crack down on one drug, trainers move on to another.

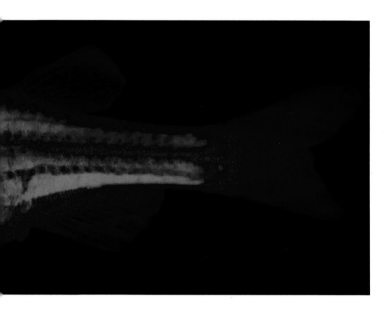

A Rescue Duo of Dog and Robot

"What is it, Lassie? Is Timmy stuck in a well? Why don't you deploy your camera-equipped robotic snake and let it do the work?" That's how an episode of *Lassie* might play in the future, thanks to scientists who have designed a dog-robot team that could one day save lives without putting the rescue animal in harm's way.

Biomimetic robots, which are designed to emulate animals, have been in development for years to serve as search-and-rescue tools. What makes the CARD (Canine Assisted Robot Deployment) system special is that it depends on a dog to operate the robot. At the scene of an earthquake or other disaster, rescuers could deploy the team to get to trapped humans much faster. Instead of waiting for a primitive robot snake to slowly slither its way through rubble, rescuers can send the dog bounding as far as it can safely go. At that point the dog is trained to bark, which signals to the robot that it's time to go into action. The tethered robot then crawls into any space the dog can't get to, providing a live video feed for rescuers waiting at a safe distance.

With only 48 hours to find disaster victims before the odds of survival drastically drop, the added speed a running dog provides could be a lifesaver.

TEAMWORK *A rescue dog being fitted with robot-deployment gear.*

Turns Out the Dingo Did It

In 1980, 9-week-old Azaria Chamberlain disappeared from her family's campsite near Ayers Rock in Australia, prompting her distraught mother to tell authorities that her daughter had been abducted by a dingo—a wild, wolflike canine that has coexisted uneasily with humans on the Australian continent for at least 3,000 years. But officials, suspecting foul play, accused Azaria's mother, Lindy Chamberlain-Creighton, of murder. In 1982 she was convicted and sent to prison. The ruling was overturned several years later, and Chamberlain's saga was made into a 1988 film, *A Cry in the Dark,* starring Meryl Streep (who uttered the immortal line "The dingo's got my baby!").

For 32 years, Azaria's fate was never conclusively solved—until June 2012, when a coroner

A DISAPPEARANCE *Michael and Lindy Chamberlain with a photo of their daughter after an inquest in 1981.*

ruled that she had in fact been abducted and killed by a dingo.

The ruling is a sign of how much more scientists understand about the dingo. Though more closely related to wolves than domestic dogs, dingoes were thought to be too small to abduct a baby. But what they lack in size, they make up for in street smarts. The Chamberlain case was the first widely reported episode of a dingo killing a small child. But three more children have been attacked in recent years—including a 9-year-old boy who was killed—and the verdict has been overwhelmingly welcomed in Australia.

3

DINGO ATTACKS AGAINST
CHILDREN IN AUSTRALIA HAVE BEEN
DOCUMENTED IN RECENT YEARS.

Planes May Soar Like an Albatross

Logging up to 50,000 miles of flight time a year and with a wingspan that can reach a dozen feet, the albatross has long been the king of effortless flight, gliding thousands of miles with barely a beat of its wings. Its food runs over the ocean can last up to 10 days, which may explain its uncanny ability to sleep while flying. For the first time, the albatross's flight acumen has been decoded by using advanced GPS receivers placed under its feathers, and the findings may influence the design of future aircraft.

The albatross uses a technique called dynamic soaring, which includes the ability to sense the smallest changes in wind direction and air pressure. The albatross follows a delicate choreography that harnesses its power. First, it glides near the water's surface, where its air speed slows. It then suddenly turns directly into the wind and allows itself to be pushed upward with increasing speed. While rocketing 45 feet skyward, the bird gains 3.6 times more speed than when it was cruising. It then turns away from the wind and descends back to the water. The result is steady, if circuitous, flying that uses very little of its own energy.

The albatross has a unique tendon in each shoulder that allows it to lock its wings in place, making it closely resemble a fixed-wing aircraft. Aeronautical engineers are taking the cue, and will try to emulate dynamic soaring to help extend the airtime for drones and other small aircraft. It's hard to imagine a 747 swooping up and down (not to mention how it would affect passengers), but engineless planes might soon be able to soar above the ocean for months, riding the currents as gracefully as an albatross.

THROUGH A METHOD CALLED DYNAMIC SOARING, THE ALBATROSS—WITH A WINGSPAN OF UP TO

12

FEET—CAN GLIDE THOUSANDS OF MILES WITHOUT FLAPPING.

Discovering the Littlest Leaper

The competition has been fierce, but a new champion has been crowned in the mini-vertebrate class: *Paedophryne amauensis,* a frog so small that it measures less than one third of an inch and can lounge comfortably on a human fingernail. With the discovery of this petite leaper, the world's former smallest frogs—the Brazilian gold frog (*Brachycephalus didactylus*) and the slightly larger Monte Iberia eleuth (*Eleutherodactylus iberia*)—have been knocked down to second and third place, respectively. And the world's former smallest vertebrate, a type of fish known as *Paedocypris progenetica,* has been stripped of its title as well. Tough break, guys.

The frog was located deep in the island of Papua New Guinea in 2009 by a team of researchers from Louisiana State University. The researchers listened to the frog's chirping sounds and discovered it dwelling in the moist leaf litter on the rainforest floor. Capable of jumping 30 times its body length, the little frog is most active at twilight, when it hunts for invertebrates.

In January 2012 the science journal *PLOS ONE* officially declared the frog the new world champ. With many amphibian species now under threat of extinction, herpetologists say the discovery is a boon for research and will help them understand how such extremely small creatures evolved.

A NEWLY DISCOVERED TYPE OF FROG MEASURES LESS THAN

ONE THIRD OF AN INCH.

BRACHIOSAURUS
With the new calculations, the Jurassic-period herbivore has shed 57 tons.

New Math Slims Down Dinosaurs

In *Jurassic Park,* they shake the earth so much they create waves in water puddles, but new research suggests that dinosaurs were lighter on their feet than previously imagined. Biologists at the University of Manchester used laser imaging and 3-D computer modeling to calculate the minimum amount of skin required to wrap around a skeleton. From that calculation, they estimated the animal's total volume. They tested the process on 14 modern-day animals, finding that they had to add about 20 percent more body mass to their calculation from the minimum skeletal "wrap" volume.

Scientists then went to Berlin's Museum für Naturkunde and applied the technique to the largest mounted dinosaur skeleton in the world, the giant *Brachiosaurus brancai.* After its 80-foot-long skeleton was measured, the *Brachiosaurus* was calculated to have shed an impressive 114,000 pounds—down from a once-estimated 80 tons to a lithe 23 tons, though that's still the rough equivalent of six elephants.

Body weight is key to understanding how an extinct animal lived—its life span, anatomy, and diet. Before supercomputers, estimating it was an inexact science, sometimes employing sculptures modeled on artists' interpretations that were then submerged in water to measure their volume. Scientists are now hopeful the animals' weight can be calculated with consistent accuracy. "I think the early estimates were set in that big, fat, and slow lizard mindset before the dinosaur renaissance," the study's author, William Sellers, told Discovery News.

One result of the new math is that many dinosaurs will be bumped down to a lighter weight class. Textbook illustrations and museum exhibits will need to be updated to reflect their svelter figures. And the next time a *T. rex* goes on a cinematic rampage, there should be a bit less rumble in the jungle.

101

Archaeology

ROMAN IMPERIAL NAVY ▪ WHO DISCOVERED FIRE? ▪ DINO-BIRD DILEMMA ▪ MYSTERIOUS TWIN LION STATUES
AN IRON AGE COIN HOARD ▪ FINDING SHAKESPEARE'S FIRST STAGE ▪ A TABLET'S LOST LANGUAGE
HAIL TO A VIKING VIP ▪ BONES OF THE MISSING KING?

Riches of the Mayan Royals

*New tomb discoveries containing the remains
of early Mesoamerican kings and queens are filling
in the blanks about their classic founding period.*

BY MICHAEL Q. BULLERDICK

One of the common complaints of the archaeologist's profession is that few discoveries without a link to the Bible can be counted on to generate lasting enthusiasm outside academia. The ancient Mayans, however, have been an exception to this rule in recent years. In fact, judging by the waves of media hype and Google trends, the Mayans are hot—Justin Bieber hot. Not since Egypt's King Tut was introduced to the world in 1922 has a topic of archaeology been the focus of such mainstream curiosity.

Unfairly to Mayans, such intense interest can be traced in large part to a major falsehood about them—in this case, their so-called Doomsday prophecy for December 21, 2012. While based on misunderstandings of Mayan calendric cycles, the resultant buzz brought a windfall of funding for Mesoamerican studies that led, sometimes inadvertently, to discoveries of several royal burial sites. Now a handful of these are filling in the blanks about dynasties around the Mayan founding period.

A wealth of evidence suggests that what is generally referred to as "the Mayan Empire" was actually a collection of city-states along the Yucatán Peninsula that were unified, if only loosely at times, by cultures with very similar beliefs. As this civilization developed during three distinct phases from 2000 B.C. to A.D. 1500, it stretched across what is now central and southern Mexico to Belize, Guatemala, Honduras, and northern El Salvador. A great deal of what we know about the Mayans comes from comprehensive examinations of their awe-inspiring stone structures—pyramids and temples rivaling those of Egypt—but the rest has come about, more directly, from the Mayans themselves. Obsessive scribes, these Mesoamerican scholars graciously detailed much of their history and royal lineage. But while troves of recovered artifacts corroborate the epigraphic

BURIAL BOUNTY *Within pyramid E in Nakum, Guatemala (above), and other previously well-explored ancient structures, archaeologists have recently stumbled upon the richly adorned tombs of several significant Mayan royals, including an unknown queen (below) whose head rested between ornately decorated sacred vessels.*

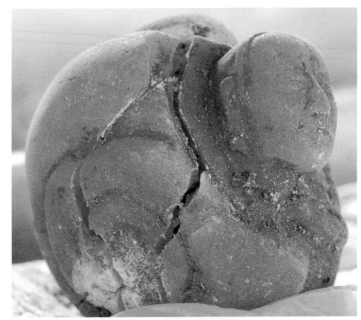

WHO'S WHO *Clues to the identity of tomb remains are often found among burial offerings: (clockwise from top) a vessel depicting Queen K'abel, a cup inscribed to an unknown prince of Uxul, and a jade vulture pendant that denotes King K'utz Chman's priestly status.*

undisturbed in cemeteries on the outskirts of town. To the contrary, they raised generational awareness and honored their ancestors in part by relocating some sacred remains to safer, more prominent "burial collectives" at the centers of thriving cities. These "upgraded interments" were constructed inside modified or new edifices built as the result of population growth and architectural advances.

A prime example of this is the one in Nakum (northeastern Guatemala) that held the nearly intact remains of an unknown Mayan queen whose skull rested between two sacred vessels. In 2011, a team from the Jagiellonian University Institute of Archaeology in Krakow, Poland, discovered the queen's 2,000-year-old burial vault after first detecting cracks in the floor of a 1,300-year-old upper vault that contained the sparse remains of a lesser royal.

The following year, researchers from the University of Bonn and Mexico's National Institute of Anthropology and History (INAH) stumbled upon a vault just five feet under the floor of a southern room in the palace at Uxul, near Campeche, Mexico. This chamber hid the entrance to a 1,300-year-old, richly adorned burial crypt containing the remains of an unknown Calakmul-dynasty prince whose skull was also covered with a unique vessel. His royal title was confirmed by the inscription on a ceramic cup that read, "This is the drinking vessel of the young man/prince."

A similar find in the Mayan capital city of Palenque, in the Mexican state of Chiapas, has yet to yield a sarcophagus, but the INAH team believes it has good reason to continue searching. Although the tomb in Temple XX was discovered in 1999, just 19 feet below the upper floor of the 60-foot-high pyramid, archaeologists were forced to wait 12 years before entering it until some much-needed structural stabilization was complete. Their leading hypothesis was that the tomb belonged to the fifth-century King K'uk' Bahlam I, the first ruler of the Palenque city-state and an ancestor of Pakal the Great, the historic emperor of the Mayan middle period. Pakal was found in 1952, buried nearby in the Temple of Inscriptions, which chronicled 200 years of the family dynasty.

Pictures of the tomb's interior that were taken using remote-controlled cameras in 1999 and 2011 paled in comparison with the actual reveal in 2012, which exposed an interior of magnificent murals that depicted the legendary nine black lords of the underworld standing out against an intense blood-red field. Bone fragments and burial offerings were also recovered, but the murals, which are similar in style and theme to Pakal's, are reason enough to suppose that a lower primary tomb containing the remains of a great royal such as K'uk' Bahlam I will eventually be located.

Far more significant than even these royal remains

records of their prime years, there had remained a disappointing lack of hard evidence to back up accounts of their pivotal classic (A.D. 250–900) and pre-classic (2000 B.C.–A.D. 250) periods.

Those gaps are beginning to narrow, thanks to a handful of tomb discoveries made deep inside structures where—oddly enough—research had been ongoing for decades. Several of these found over the last six years contained the remarkable remains of previously elusive, though well-sourced, Mayan nobility whose vaults were often well concealed below more recently constructed ones that contained separate remains. These "secondary" individuals sometimes turned out to be genealogically related to the more deeply buried "primary" subjects, but most often they belonged to unrelated people thought to be sacrificial victims. Moreover, evidence from tombs of this type often suggested that the contents may have been relocated there. In fact, the resulting inconsistencies of several remains in terms of age, gender, skeletal completeness, and articulation can be accurately described as a kind of ancient mash-up.

At first glance, all this mixing and moving of bones seems profoundly ghoulish, but the Mayans did not fully embrace our modern belief that the dead should remain

DIGGING IN *Archaeologist Olivia Navarro-Farr works carefully to expose the headdress of the seventh-century warrior queen K'abel.*

Mayans honored their ancestors by moving sacred remains to safer, more prominent "burial collectives."

were the ones from dual tomb discoveries in June 2012 that stunned archaeologists. The first was of a tomb in the royal city of El Perú-Waká in northwestern Guatemala containing the remains of K'abel, the legendary warrior queen ("Kaloomte") of the classic era. Typically, the concealed tomb rested just 22 feet below the surface of another vault within a previously well-explored pyramid. A team from Washington University in St. Louis stumbled upon the tomb after deciding to focus on religious shrines. Although the remains are in poor condition, the key to identifying them came from among the vast burial offerings—a small alabaster jar carved in the shape of a conch shell detailed with the head and arms of a prominent woman and inscribed with both of K'abel's monikers: "Lady Water Lily Hand" and "Lady Snake Lord." Additionally, anthropologists agree that the shape of the skull matches the queen's visage on plaques referring to her and her husband, King K'inich Bahlam II.

While such a find is hard to top, the second of the June

2012 tomb discoveries was momentous, despite its current lack of remains, for being the oldest ever discovered: the tomb of King K'utz Chman, who led the transition from pre-Mayan Olmec rule to an eventual Mayan dominance that began with his emphasis on written historical records and pyramid building. Inside the tomb, located at a dig in western Guatemala, local archaeologists have discovered burial jewels that included a jade pendant of a vulture, an apparent identifying marker. Carbon dating places the burial at the pre-classic Mayan period, somewhere around 770–510 B.C.

Undoubtedly, important data from the recent roundup of ancient royal tombs will aid archaeologists immeasurably during future explorations of the Mayan city of Holtun near Petén, Guatemala, which satellite 3-D imaging technology revealed under thick jungle in 2012. But the takeaway for known Mayan sites is equally compelling: even the most extensively explored and well-documented ruins may still hold a secret or two buried deeply, one below the other, within their cold stone interiors. Granted, this will require more funding and even more time. But time, at least, is something we still have in abundance—even if, during the run-up to December 21, 2012, many people around the world suspected otherwise.

Where the Romans Built Grand Ships

For about a century, archaeologists have struggled to locate the Roman Imperial Navy's primary shipyard. A few notable candidate sites have been found—Monte Testaccio on the Tiber River near Rome, and another at neighboring Ostia—but none has been impressive enough to fit the bill of "fleet headquarters." Until now, that is.

In September 2011 archaeologists from the University of Southampton and the British School at Rome, who had been excavating a site at Portus, Italy, for more than a decade, uncovered a massive rectangular building that was erected in line with a nearby hexagonal harbor basin. The crumbling ruin, which dates to the second century, runs along the basin's side and rests on thick, brick-faced concrete piers. The enormous structure—919 feet long, 190 feet wide, and six stories high—contains multiple bays that would have been used to build, repair, and house ships of up to at least 350 tons.

Beyond its grand scale, what makes the case for the Portus site as Roman Navy central is its location within a city that was a key maritime port even before the shipyard's construction between A.D. 110 and 117. Portus sits where the Tiber meets the sea, making it an ideal location for imports, troop transport, and a naval defense of Rome, just 20 miles away. Such a distance would have been convenient for seabound nobility, and the adjoining opulent palace unearthed in 2009 supports this theory. Inscriptions at Portus refer to a guild of prominent shipbuilders, and a mosaic that originated near Portus depicts a building with ships in each bay and a façade that is similar to that of the Portus structure.

Soon archaeologists will search for ramps used to launch the huge Roman ships, but it's likely they decayed centuries ago. Researchers will also look for clues to the date and cause of the shipyard's ultimate decline.

IN DRY DOCK *A computer rendering shows Portus as it would have looked after being built during the reign of Emperor Trajan (circa A.D. 98–117).*

Figuring Out Fire: Did It Happen 300,000 Years Earlier?

Fire—its discovery and controlled application—was an evolutionary game-changer for early humans. The prevailing theory places our mastery of it at roughly 790,000 years ago, but a recent analysis of deposits discovered in South Africa indicates this crucial event may have originated 300,000 years earlier, by a more ancient species of man known as *Homo erectus*. The deposits from Wonderwerk Cave, a site near the edge of the Kalahari Desert in Northern Cape province, included ash from burned grass, leaves, brush, and animal-bone fragments as well as flint and stone tools. Microscopic examination indicates they were burned at temperatures between 400 and 700 degrees Celsius—consistent with maximum heat output for the natural fuels and common for small campfires. Previous "earlier" finds—in Swartkrans, South Africa; Chesowanja, Kenya; and Gesher Benot Ya'aqov, Israel—had been ruled inconclusive because they were situated out in the open, where they may have been started by lightning, or because they were found in caves filled with bat droppings, a combustible material. But the compelling evidence from Wonderwerk, an "indoor" setting that shows no significant quantities of bat guano, comes closest to satisfying the requirements for evidence that humans were capable of creating fire as opposed to simply transporting it.

Dino-Bird Ruffles Feathers

The theory that modern-day birds evolved from dinosaurs has held sway for over 150 years and rests on a handful of well-studied fossil specimens originating in Eastern and Central Europe and in Asia. Chief among these is *Archaeopteryx*, discovered in 1861 in Bavaria, Germany. Although clearly a dinosaur, *Archaeopteryx* possesses many anatomical features common to birds, including feathers, which elevated it to the iconic status of "proto-bird." But now a much older specimen is challenging that status while raising questions about just what constitutes a "proto-bird."

All of this began with the discovery in 2011 of *Xiaotingia zhengi,* a fossilized dino-bird in Liaoning Province, China, that was the size of a chicken. In addition to the similarity in size to its hallowed fossilized friend, *Xiaotingia zhengi* exhibits dual anatomic traits: birdlike feathers, a wishbone, three-toed feet with long middle bones, and long forearms (perhaps allowing it to fly), mixed with a dino-like elongated skull, sharp teeth, and forearms ending in claws. These similarities, however, are a big problem for both proto-bird pretenders. Given that *Archaeopteryx* lived 150 million years closer to the rise of birds on earth, it should have evolved to a far more avian state than its predecessor, *Xiaotingia zhengi*. That it didn't places both creatures closer to the dinosaur side of the evolutionary timeline than to the critical proto-bird point. Computer modeling backs this up, fitting them both squarely in the Deinonychosauria branch of the late Jurassic and Cretaceous periods that also includes velociraptors. This classification effectively widens the gap between dinosaurs and birds.

ANGRY BIRDS *Experts believe* Xiaotingia zhengi *may have looked like a bluejay on steroids.*

GROUNDED: RECENT STUDIES SHOW THAT DESPITE

HAVING "WINGS,"

SO-CALLED PROTO-BIRDS LACKED FEATHERS STRONG ENOUGH FOR LIFTOFF OR SUSTAINED FLIGHT.

Twin Lion Statues of Turkey Retain Their Secrets

One of the greatest challenges that archaeologists face is getting to a rumored site or accidental discovery ahead of professional looters. That was the case in 2008 when rumors began to circulate about the discovery in Karakiz, Turkey, of twin lion statues carved out of monolithic granite. Local tomb raiders used dynamite to blast the giant markers in two and then dug below them and in the surrounding fields in search of the buried treasure they believed the lions were meant to guard. The looters walked away empty-handed, but the destruction they left behind is making it harder for archaeologists who, after finally securing funding, made it to the site in 2011 to properly excavate it and answer the prevailing questions: Why lions? Why there? And what were the markers intended to call out?

Carbon-dating techniques and the style of workmanship indicate that an Indo-European people known as the Hittites sculpted the statues around 1400–1200 B.C., according to archaeologists Geoffrey Summers and Erol Özen. Although similar in form and pose, the five-ton lions are stylistically different, a clear indication that they were sculpted by two different artists. After careful analysis, archaeologists and preservationists were able to cement the cracked pieces back onto the still-standing portions of the monoliths. Each revealed a detailed depiction of a lion captured in proud stride: its profiled form, bone structure, and forward-leaning gait. At first it even appeared as if the lions had been painted or stained to reflect the sandy color of lion fur, but closer examination proved the coloration was the result of centuries of oxidation.

Lions, including the Asiatic species depicted on the granite slabs, do not roam present-day Turkey, but the regal beast was well known and respected by the ancient people who inhabited Turkey and other parts of the Middle East. If the Hittites meant for the lions to guard a settlement, city, or palace, however, none has so far been found. Given their collective mass, it's not likely the statues were carved and transported to their current locations. One theory is that the lions are markers for now-underground wellsprings that the Hittites considered holy, purifying places worthy of such adornments. A stone basin measuring seven feet in diameter that was found near the lions seems to support this point. Future excavations will be undertaken to explore the theory.

LION'S DEN *Lions, wellsprings, and the mountains near the statues were considered sacred by the Hittites.*

THE HITTITES WERE AN ANATOLIAN PEOPLE WHO SPOKE AN INDO-EUROPEAN TONGUE. THEIR CULTURE'S ZENITH WAS DURING THE 14TH CENTURY B.C.

SERIOUS COIN *X-rays of the fused block reveal that it also contains some gold and silver articles.*

The Trove Forever Hidden From Caesar

Anyone who has ever tried out a metal detector has fantasized, if only briefly, about discovering buried treasure. For the persistent treasure-seeking duo of Reg Mead and Richard Miles, that dream finally became real after 30 years of traversing the same farm fields. And the treasure proved to be a major archaeological find: the largest hoard of Late Iron Age coins ever unearthed. The duo's stunning June 2012 find was located on Jersey, the largest of the Channel Islands, and consists of 50,000 silver and bronze coins. Early evidence suggests the coins date from the Iron Age, around 50 B.C., and originated in Armorica (Brittany and Normandy) but were probably buried for safekeeping from Julius Caesar's advancing army. A team of archaeologists and metallurgists is handling the painstaking chore of separating the coins that, over the centuries, had oxidized and been compressed into a solid, three-quarter-ton block. Estimates put the value of individual coins at about $200, depending on condition, and the total estimate in the neighborhood of $12 million to $16 million. But the coins will not be sold—at least not in the short term. Questions of legal ownership have been raised because the Isle of Jersey, a British crown dependency, falls outside the reach of the Treasure Act, which places such finds in a government trust. Jersey's treasure laws are based on more complicated medieval property laws.

THE HOARD OF 50,000 COINS
WAS FOUND OXIDIZED AND FUSED TOGETHER INTO A GIANT BLOCK WEIGHING THREE QUARTERS OF A TON.

Shakespeare's Curtain Rises

"All the world's a stage," wrote Shakespeare. But where in the world is the one he and his players first strode upon? The site has eluded archaeologists for a century, so its accidental discovery in June 2012—as if on cue for worldwide Shakespeare festivities already underway—represented a plot device worthy of the bard himself. The site was discovered by a construction crew working on a Hewett Street community-redevelopment project in Shoreditch, east of London's business district. Although Shakespeare is properly associated with the Globe Theatre, he spent some of his early professional years (1597–99) at a tiny round playhouse while the Globe was under construction. Much of the venue has survived intact; however, the position of two surviving gallery walls indicates that the famed stage is currently located under the foundation of a modern tavern. Upon its completion in 1577, the rough-hewn theater was formally named the Curtain after the long road that fronted it. Contemporary accounts, however, reveal it was referred to more frequently as the "Wooden O" after a characterizing line from *Henry V*, which premiered there. *Romeo and Juliet* opened there too, followed by works from other notable playwrights before the theater went into decline. Its exact location was lost to history somewhere between 1622 and 1640.

DOWNSTAGE
Shakespeare's Curtain Theatre was located just five feet below modern street grade. His first audiences, peasant "groundlings," would have paid a penny per head to be entertained there.

TALKING TURKEY *The tablet that bears the unknown language was found at the ruins of a regional governor's palace in Ziyaret Tepe, Turkey.*

TOOLS OF THE TRADE *Although they date to 1,000 years ago, the Viking chieftain's rust-encrusted iron sword, ax head, and brooch pin display a remarkable level of metallurgy.*

A Clay Tablet's Mysterious Language

Almost 1,000 years before Christ, the Assyrian city of Tushan—part of an empire stretching from modern-day Iraq and Iran through southeastern Turkey—was a bustling, ethnically diverse center for trade. More than a dozen known languages were spoken, and of these only one remains silent: that of the Assyrian-conquered Zagros Mountain people. But a clay tablet unearthed in May 2012 in Ziyaret Tepe, Turkey, may give voice to the Zagros people once again. If that's the case, it's a happy accident: like a kiln-fired clay pot, the tablet was protected against the ravages of time after being hardened in a fire near the end of the eighth century B.C.

Once part of a royal archive, the earthen tablet is inscribed using cuneiform characters. Its content is a partial list of names along with what appears to be an incomplete chronicle of the Zagros people's forced relocation from their home in the mountains of Iran to southeastern Turkey. Archaeologist and project epigrapher John MacGinnis of England's McDonald Institute for Archaeological Research has identified approximately 60 contemporaneous names out of the 144 listed on the tablet. Seventy-five percent of the tablet's content, however, is documented in the unknown language. MacGinnis's analysis will continue, but initial findings favor the Zagros hypothesis while ruling out other contemporary languages—from Mushkian, an equally obscure tongue, to Shubrian, the most common.

CUNEIFORM, ONE OF THE EARLIEST FORMS OF WRITING, WAS INVENTED BY THE SUMERIANS (30TH CENTURY B.C.) AND MODIFIED BY THE ASSYRIANS.

Burial Boat for a Viking VIP

Unique to a Viking's view of the afterlife was a belief in Valhalla, where those who died in battle were rewarded with further glorious battle. Such beliefs were no doubt held by the Viking chieftain whose remains and burial boat were recently discovered in Scotland. Arranged around him were the tools of his "pious" trade—sword and hilt, ax and shield, spear, sharpening stone, food pots, and a drinking horn—proving he was suited up and ready to play.

The spectacular discovery, the first on the British mainland, was made in 2011 on the Ardnamurchan Peninsula, a landing spot on the primary north-south sailing route between Ireland and Norway at the time the chieftain met his end approximately 1,000 years ago. Just how he died will likely remain a mystery, since only partial remains were present within the boat's detailed earthen impression, but there is enough evidence for archaeologists to form a picture of the ancient warrior's life. That he was buried in ceremonial fashion and laid to rest with his possessions—items that would have been extremely valuable to others—solidifies his status as an important man or chieftain. Radioactive isotopes from the Viking's bones should determine more about his age, origins, and genetics, and an analysis of wood slivers should reveal the species of trees the boat was made from and where it may have been constructed. Additional analysis of the artifacts, including hundreds of metal boat rivets and dozens of iron pieces, will provide supporting data about the Vikings' craftsmanship and their lives in general.

Are These Richard's Bones?

Few historical figures have been as maligned as England's King Richard III. But if an analysis of a skeleton discovered in 2012 under a parking lot in Leicester, England, proves it to be the missing king, members of the Richard III Society hope renewed curiosity about him will help restore his reputation. No less a dramatist than Shakespeare is responsible for expanding the view of a ruthless, grotesque Richard that had become conventional wisdom between the time of the king's death in 1485 and the writing of *Richard III* in approximately 1591. But history is also influenced by the victors—in this case the Tudor family, whose killing of Richard, the last of the Plantagenet kings, won them the crown in the Wars of the Roses. The Leicester remains are those of an adult male of Richard's age (about 32) at the time he died. The back of the skull reveals it was cleaved with a bladed instrument, which accords with written reports about the manner of his death. An examination of the spine asserts the strongest link: an obvious curvature known as scoliosis, which the Tudors exaggerated mercilessly. The remains were uncovered in the choir chamber of the Friars Manor ruins, the spot listed as Richard's burial site in a prominent period report. Researchers plan to compare mitochondrial DNA from the remains with DNA from one of the king's descendants to see if they can confirm a match.

IN THE DEEP BOSOM … *The precise location of the Friars Manor ruins was unknown until 2012.*

IT JUST AIN'T SO …

GARDEN OF STONE *Urn after urn at Carthage contains the cremated remains from multiple children of varying ages from several different families.*

Child Sacrifice Debunked

Among the world's more gruesome tourist attractions is the Tophet of Carthage (in modern-day Tunisia), an ancient burial ground containing the charred remains of 20,000 infants interred there between 730 and 143 B.C. Not a trace of adult remains has ever been found in the site's acre-size field, just infants, young goats, and lambs—which has led to its reputation, beginning with anti-Carthaginian propaganda from Greek and Roman enemies, as a place of barbaric, ritual child sacrifice.

Now interpretations of the Carthage curiosity are becoming far less ghastly, thanks to Jeffrey Schwartz of the University of Pittsburgh, whose study of the site was published in *Antiquity* in 2011. Schwartz asserts the Tophet was merely a cemetery reserved for infants who died mostly of natural causes and were buried there after cremation. Carthaginians, it seems, did not view children under the age of 2 as full-fledged people because they were incapable of holding religious beliefs. And if infants were not worthy of adult-style burial, Schwartz postulates, how could they be deemed worthy of sacrifice? He argues that the preponderance of prenatal evidence found at the Tophet indicates that the deaths cannot be attributed to live sacrifice. Schwartz examined 348 urns containing 540 individuals from the Tophet. By measuring cranium sizes and tooth development, he found that nearly 20 percent of the infant Carthaginians died as the result of spontaneous abortion or from stillbirth. As for the goats and lambs, Carthaginians routinely offered these prize animals to their god Ba'al Hammon, including them to ensure a loved one's easy passage into the afterlife or to seal a prayer for the next child to be granted a longer life.